Stranded!

Building Math

Integrating Algebra & Engineering

The *Building Math* project was funded by a grant from the GE Foundation, a philanthropic organization of the General Electric Company that works to strengthen educational access, equity, and quality for disadvantaged youth globally.

The interpretations and conclusions contained in this book are those of the authors and do not represent the views of the GE Foundation.

1 2 3 4 5 6 7 8 9 10
ISBN 978-0-8251-6877-2
J. Weston Walch, Publisher
40 Walch Drive
Portland, ME 04103
www.walch.com
Printed in the United States of America

TABLE OF CONTENTS

INTRODUCTION

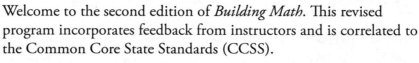

Welcome to the second edition of *Building Math*. This revised program incorporates feedback from instructors and is correlated to the Common Core State Standards (CCSS).

Building Math is a unique program that integrates real-world math and engineering design concepts with adventurous scenarios that draw students in. The teacher-tested, research-based activities in this program enforce critical thinking skills, teamwork, and problem solving, while bringing students' classroom experiences in line with the Common Core.

Each set of three activities (Design Challenges) forms one unit. The unit's activities are embedded within an engaging fictional situation, providing meaningful contexts for students as they use the engineering design process and mathematical investigations to solve problems. There are three units, and each unit takes about three weeks of class time to implement.

WHAT'S INCLUDED IN THE BUILDING MATH PROGRAM

The instructional materials include reproducible student pages, teacher pages, a DVD of classroom videos for teacher professional development and a Java applet used as a computer model in one of the activities, and a poster showing the engineering design process. The full program is also provided as an interactive PDF on an accompanying CD. This includes all of the content from the book as well as expanded CCSS correlations. The CD is intended to facilitate projecting materials in the classroom and/or printing student pages. You can also print out transparencies from the instructional pages if it suits your needs.

WHY IS ENGINEERING EDUCATION IMPORTANT?

The United States is faced with the challenge of increasing the workforce in quantitative fields (e.g., engineering, science, technology, and math). Schools and teachers play a pivotal role in this challenge. Currently, many students (mostly from underrepresented groups) are not graduating from high school with the necessary math skills to continue studies in college in these quantitative fields. Colleges and industries such as Tufts University and General Electric realize that more active participation in the pre-K–12 system is needed, and have put together this innovative program to help teachers increase math and engineering content in the middle-school curriculum.

(*continued*)

Algebra and engineering are critical fields that are worth combining. Algebraic reasoning acts as a foundation for higher levels of math learning in secondary and tertiary education, and introducing students to engineering is a way to show them how math is used as a discipline of study and a career path.

BACKGROUND

Building Math was a three-year project funded through the GE Foundation in partnership with the Museum of Science in Boston. One goal of the project was to provide professional development for middle-school teachers in math and engineering, and to explore alternative teaching methods aimed at improving eighth grade students' achievement in algebra and technology. Another project goal was to develop standards-based activities that integrate algebra and engineering using a hands-on, problem-solving, and cooperative-learning approach.

The resulting design challenges were tested by teachers in ten Massachusetts schools that varied in type (public, charter, and independent), location (urban, suburban, rural), and student demographics. Subsequently, hundreds of teachers all over the country used the materials with their students. Their experiences have informed this second edition of *Building Math,* which includes correlations to Common Core State Standards. Further, this edition includes specific, optional suggestions to allow teachers to address additional aspects of the CCSS.

PARTICIPATING RESEARCHERS AND SCHOOLS

Project Investigators: Dr. Peter Y. Wong and Dr. Bárbara M. Brizuela
Project Coordinators: Lori A. Weiss and Wendy Huang
Pilot Schools and Teachers:
- Ferryway School (Malden, MA): Suzanne Collins, Julie Jones
- East Somerville Community School (Somerville, MA): Jack O'Keefe, Mary McClellan, Barbara Vozella
- West Somerville Neighborhood School (Somerville, MA): Colleen Murphy
- Breed Middle School (Lynn, MA): Maurice Twomey, Kathleen White
- Fay School (Southborough, MA): Christopher Hartmann
- The Carroll School (Lincoln, MA): Todd Bearson
- Knox Trail Junior High School (Spencer, MA): Gayle Roach
- Community Charter School of Cambridge (Cambridge, MA): Frances Tee
- Mystic Valley Regional Charter School (Malden, MA): Joseph McMullin
- Pierce Middle School (Milton, MA): Nancy Mikels

ORGANIZATION AND STRUCTURE OF *STRANDED!*

Throughout this reproducible book, you will find teacher guide pages followed by one or more student activity pages. Each page is labeled as either a student page or a teacher page.

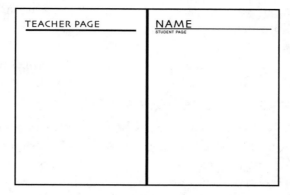

Answers to all activities and discussion questions are found in the answer key at the back of the book.

TIPS, EXTENSIONS, OPTIONAL CCSS ENHANCEMENTS, and **ASSESSMENTS** are labeled in gray boxes.

 INTERESTING INFO is provided in white boxes. This provides additional related information and resources that you may want to share with students.

The following labels are used to indicate whether you will be addressing the whole class, teams, individuals, or pairs:

| CLASS | INDIVIDUALS | TEAMS | PAIRS |

 ### THE ENGINEERING DESIGN PROCESS
Activities align with the eight-step engineering design process. See page 18 for a lesson plan to introduce the steps.

TABLES AND GRAPHS

Tables and graphs are numbered according to their order of appearance in each design challenge. Those beginning with 1 correspond to Design Challenge 1. Those beginning with 2 correspond to Design Challenge 2. Those beginning with 3 correspond to Design Challenge 3.

BUILDING MATH: PEDAGOGICAL APPROACH, GOALS, AND METHODS

The approach of the *Building Math* program is to engage students in active learning through hands-on, team-based engineering projects that make learning math meaningful to the students (Goldman & Knudson).

The goal of *Building Math* is to encourage the development of conceptual, critical, and creative-thinking processes, as well as social skills, including cooperation, sharing, and negotiation by exhibiting four distinctive methods (Johnson, 1997).

1. **Contextual learning**—Each *Building Math* book involves students in a story line based on real-life situations that pose fictional, but authentic, design challenges. Design contexts invite students to bring ideas, practices, and knowledge from their everyday lives to classroom work. Students apply math skills and knowledge in meaningful ways by using math analysis to connect inquiry-based investigations with creating design solutions.

2. **Peer-based learning**—Students work together throughout the design process. The key to peer-based learning is high amounts of productive, on-task verbalization. By verbalizing their thoughts, students listen to their own and others' thinking. This allows them to evaluate and modify one another's thinking and defend their own ideas. Verbalization also contributes to more precise thinking, especially when teachers use effective questioning techniques to ask students to explain and analyze their and others' reasoning.

3. **Activity-based practice**—*Building Math* uses design challenges (design and construction of a product or process that solves a problem) to focus peer-based learning. Students conduct experiments and systematic investigations; use measuring instruments; carefully observe results; gather, summarize, and display data; build physical models; and analyze costs and trade-offs (Richards, 2005).

4. **Reflective practice**—*Building Math* activities include questions, rubrics, and self-assessment checklists for students to document and reflect on their work throughout each design stage. Teams summarize and present their design solutions to the class, and receive and offer feedback on others' solutions.

REFERENCES

Goldman, S., & Knudsen, J. Learning sciences research and wide-spread school change: Issues from the field. Paper presented at the International Conference of the Learning Sciences (ICLS).

Johnson, S. D. 1997. Learning technological concepts and developing intellectual skills. International Journal of Technology and Design Education, 7, 161–180.

Richards, L. G. 2005. Getting them early: Teaching engineering design in middle-schools. Paper presented at the National Collegiate Inventors & Innovators Alliance (NCIIA). http://www.nciia.org/conf05/cd/supplemental/richards1.pdf

ENDURING UNDERSTANDINGS

In their book, *Understanding by Design,* Wiggins and McTighe advocate that curricula can be built by identifying enduring understandings. An enduring understanding is a big idea that resides at the heart of a discipline, has lasting value outside of the classroom, and requires uncovering of abstract or often misunderstood ideas. The list below details the enduring understandings addressed in *Building Math*. Teachers can consider these to be the ultimate learning goals for the *Building Math* series.

ENGINEERING AND TECHNOLOGICAL LITERACY

1. Technology consists of products, systems, and processes by which humans modify nature to solve problems and meet needs.

2. Design is a creative planning process that leads to other useful products and systems.

3. There is no perfect design.

4. Requirements for design are made up of criteria and constraints.

5. Design involves a set of steps, which can be performed in different sequences and repeated as needed.

6. Successful design solutions are often based on research, which may include systematic experimentation, a trial-and-error process, or transferring existing solutions done by others.

7. Prototypes are working models that can later be improved to become valuable products. Engineers build prototypes to experiment with different solutions for less cost and time than it would take to build full-scale products.

8. Trade-off is a decision process recognizing the need for careful compromises among competing factors.

MATH

1. Math plays a key role in creating technology solutions to meet needs.

2. Mathematical models can represent physical phenomena.

3. Patterns can be represented in different forms using tables, graphs, and symbols.

4. Graphs are useful to visually show the relationship between two variables.

5. Measurement data are approximated values due to tool imprecision and human error.

6. Repeated trials and averages can build one's confidence in measurement data.

7. Mathematical analysis can lead to conclusions to help one make design decisions to successfully meet criteria and constraints.

8. Analyzing data can reveal possible relationships between variables, and support predictions and conjectures.

REFERENCES

International Technology and Engineering Educators Association (ITEEA). 2007. *Standards for Technological Literacy: Content for the Study of Technology.* (Third ed.). Virginia.

Wiggins, G., and J. McTighe. 2005. *Understanding by Design.* (2nd ed.). Prentice Hall.

STRANDED! OVERVIEW: STORY LINE AND LEARNING OBJECTIVES

	DESIGN CHALLENGE OVERVIEW	STUDENTS WILL:
WHERE ARE WE?	Students imagine that they are on a school field trip to New Zealand to see where The Lord of the Rings movies were filmed. The plane is on course for New Zealand until a severe thunderstorm causes both the engine and radio to fail. The plane is forced to crash down somewhere in the South Pacific. The students climb into the emergency raft and drift in the ocean. Eventually, they wash ashore on what seems to be a deserted island. Students will use airplane speeds, flight durations, and a drawn-to-scale map to determine their approximate location in the South Pacific.	• Interpret a scale on a map. • Use proportional reasoning to calculate actual distance and drawn distance on a map according to a scale. • Use the relationship speed = distance/time to find one quantity given the other two quantities. • Solve a multistep problem. • Use a ruler.
DC 1: A STORM IS APPROACHING!	A severe thunderstorm is heading toward the island and will be arriving in just a few hours. Students must build a shelter to protect them from the rain and strong winds. There is a limited supply of materials on the island. Students will work in teams to design and build a scale model of a shelter that can withstand strong winds, is water resistant, and provides a minimum of 1 cubic meter of personal space for each member of the team.	• Identify similar three-dimensional objects. • Identify corresponding dimensions of similar objects. • Use a ruler to measure three-dimensional objects. • Calculate surface area and volume of rectangular prisms. • Analyze a table of values for patterns. • Generalize patterns using symbols. • Use a scale to calculate the amount of materials available for building a scale model. • Apply the engineering design process to solve a problem.
DC 2: WE NEED WATER!	The average person cannot survive for more than one week without fresh water. Ocean salt water surrounds the island, and there is no supply of fresh water on the island. Students will have to rely on rainwater for survival. In this activity, students discover an irregularly shaped piece of plane siding. Students will use this piece of siding to create a rainwater collector. The challenge is to create a collector design with a large volume as well as functionality. Students will investigate the relationship among height, radius, surface area, and volume of a cylinder to help them with their design.	• Find the area of an irregular two-dimensional shape using strategies for finding the areas of triangles, rectangles, and parallelograms. • Use a ruler to measure three-dimensional objects (cylinders and rectangular prisms). • Calculate the surface area and volume of three-dimensional objects. • Analyze a table of values for patterns. • Make and test conjectures about the relationship between surface area and volume, and dimensions and volume. • Produce and analyze line graphs that represent the relationship between two variables. • Apply the engineering design process to solve a problem.

DC 3: BALANCING ACT!

The students are in luck! Some Maori people, canoeing nearby, have spotted the students and have come to take them to the mainland of New Zealand. However, the canoe is extremely unstable. If the canoe is even the slightest bit unbalanced, it will tip over. Students will work in teams to learn how to balance multiple objects of different weights on a seesaw-like platform. Then they will design and test a loading plan in order to get everything, including themselves, into the canoe without it tipping over.

- Investigate how the weight and distance of objects on a horizontal platform with a center fulcrum relate physically and mathematically to keep the platform balanced.

- Generalize and represent a pattern using symbols.

- Apply the engineering design process to solve a problem.

ASSESSMENT OPPORTUNITIES AND MATERIALS LISTS

The tables on the next two pages list the opportunities for formative assessment by using rubrics, probing students' thinking during class time, and reviewing responses to certain questions. The tables also show the materials needed for each design challenge. The numbers in the assessment column refer to the steps of the engineering design process.

	Assessment	Materials
WHERE ARE WE?	• See assessment math problem on page 14.	• map on transparency • blank transparencies • transparency marker • overhead projector
DESIGN CHALLENGE 1: A STORM IS APPROACHING!	• **2. Research:** Assess whether students can use the scale factor to calculate the amount of material they will use to build their model. • **4. Choose:** Assess engineering drawing based on quality and communication. Use the rubric on page 128. • **5. Build:** Assess model/prototype artifact based on craftsmanship and completeness using the rubric on page 131. • **6–8. Test, Communicate, Redesign:** Assess written responses and student observations during test, communicate, and redesign steps based on model performance, completeness, and quality of reflection. Use the rubric on page 132. • **Individual Self-Assessment Rubric:** Students can use the checklist on page 50 to determine how well they met behavior and work expectations. • **Team Evaluation:** Students can complete the questions on page 52 to reflect on how well they worked in teams and celebrate successes, as well as make plans to improve teamwork. • **Student Participation Rubric:** Make copies of the rubric on page 133 to score each student's participation in the design challenge.	For each team: • 1 calculator • 20 craft sticks (12 cm) • aluminum foil (10 cm × 16 cm) • wax paper (12 cm × 20 cm) • string (24 cm) • clay (64 cm³) • 1 sheet transparency • 1 set transparency markers (4 colors) • ruler (metric) For the class: • spray bottle with water • stiff cardboard at least 21 cm × 26 cm • cord stock (several sheets for making nets)
DESIGN CHALLENGE 2: WE NEED WATER!	• **2. Research:** Assess whether students are measuring accurately, are able to correctly calculate total surface area of a square prism, and are able to calculate the volume of a square prism. • **2. Research:** Assess whether students can generalize the rule for creating a square prism with the largest volume given a fixed amount of surface material. • **2. Research:** Assess whether students recognize the similarity in the maximize volume rule for the cylinder as for the square prism. • **2. Research:** Assess whether students can apply the rule to maximize volume in cylinders and square prisms to other kinds of prisms.	For each team: • 1 calculator • 1 cylinder (can) from Appendix B • 1 cylinder (can) from Appendix C (optional) • 1 square prism (box) from Appendix D • poster board to make above nets and plane siding • 1 ruler • 1 roll tape (invisible) • 1 pair scissors • dry navy beans (about 0.4 L) • colored pencils or markers (2 different colors)

	Assessment	Materials
DESIGN CHALLENGE 2: WE NEED WATER! (CONT.)	• **4. Choose:** Assess engineering drawing based on quality and communication. Use the rubric on page 129. • **4. Choose:** Assess whether students can calculate the surface area of their design. • **5. Build:** Assess model/prototype artifact based on craftsmanship and completeness. Use the rubric on page 131. • **6–8. Test, Communicate, Redesign:** Assess written responses and student observations during test, communicate, and redesign steps based on model performance, completeness, and quality of reflection. Use the rubric on page 132. • **Individual Self-Assessment Rubric:** Students can use the checklist on page 91 to determine how well they met behavior and work expectations. • **Team Evaluation:** Students can complete the questions on page 93 to reflect on how well they worked in teams and celebrate successes, as well as make plans to improve teamwork. • **Student Participation Rubric:** Make copies of the rubric on page 133 to score each student's participation in the design challenge.	For each team *(continued)*: • 1 graduated measuring cylinder or water bottle with mL markings • 1 funnel (can be made from poster board rolled up in a cone with tip cut off) • 2 pieces of Velcro fasteners • 4 pieces of string or yarn (7–12 cm) For the class: • 1 sheet of letter-sized transparency cut in half • additional dry beans (including the ones given to the groups, should total at least 5L) • transparencies • large flat box, such as a file box lid
DESIGN CHALLENGE 3: BALANCING ACT!	• **2. Research:** Assess how students arrive at correct solutions (e.g., trial and error, generalizing the pattern they described to more than two objects, and so forth). • **2. Research:** Assess whether students can use pictures or symbols to describe the general pattern or rule that they found from the balancing activity. • **3. Brainstorm:** Assess whether students can use the rule to balance items and people described in constraints of the problem. • **4. Choose:** Assess engineering drawing based on quality and communication. Use the rubric on page 130. • **6–8. Test, Communicate, Redesign:** Assess written responses and student observations during test, communicate, and redesign steps based on model performance, completeness, and quality of reflection. Use the rubric on page 132. • **Individual Self-Assessment Rubric:** Students can use the checklist on page 121 to determine how well they met behavior and work expectations. • **Team Evaluation:** Students can complete the questions on page 123 to reflect on how well they worked in teams and celebrate successes as well as make plans to improve teamwork. • **Student Participation Rubric:** Make copies of the rubric on page 133 to score each student's participation in the design challenge.	For each team: • 1 Math Balance kit • 1 metric ruler For the class (optional): • transparency markers • transparencies • overhead projector

STRANDED! MASTER MATERIALS LIST

Qty	Item	C* or R*	1. A Storm Is Approaching!	2. We Need Water!	3. Balancing Act!
PER GROUP					
1	calculator	R	✓	✓	
2	different-colored pencils or markers	R		✓	
1	dry navy beans (400 cm³, or 400 mL)	R		✓	
1	funnel (or rolled-up poster board)	R		✓	
1	graduated measuring cylinder (or 500-mL water bottle) with mL markings	R		✓	
1	Math Balance kit[1]	R			✓
1	pack of transparency markers (4 colors)	R	✓	✓	✓
1	pair of scissors	R		✓	
4	pieces of yarn or string (about 7–12 cm per piece)	R		✓	
1	plastic quart container	R		✓	
1	ruler (metric)	R	✓	✓	✓
2	Velcro fasteners (about 1.25 cm each)	R		✓	
1	clay (64 cm³)	C	✓		
20	craft sticks (12 cm)	C	✓		✓
1	piece of aluminum foil (10 cm × 16 cm)	C	✓		
4	pieces of poster board (56 cm × 71 cm)	C	✓	✓	
1	piece of wax paper (12 cm × 20 cm)	C	✓		
1	roll of tape (invisible)	C		✓	
1	string (at least 250 cm to be cut into pieces later)	C	✓	✓	
3	transparencies (standard letter size)	C	✓	✓	✓
PER TEACHER					
1	dry navy beans (additional to teams for a total minimum of 5 L)	R		✓	
1	large flat box, such as a file box lid	R	✓		
1	overhead projector	R	✓	✓	✓
1	piece of cardboard (stiff, at least 21 cm × 26 cm)	R	✓		
1	spray bottle with water	R	✓		
1	pack of card stock (standard letter size)	C	✓		
1	pack of chart paper	C	✓	✓	✓

*C = Consumable
*R = Reusable

[1]The Math Balance kit can be ordered from Classroom Products Warehouse http://classroomproductswarehouse.com, item CN5532020, for $13.99 per kit. Several activities are available online to extend the use of the balance kits beyond the one activity in the *Building Math* program. When activities require more weights than are included in the kit, have teams combine to share kits and weights.

COMMON CORE AND ITEEA STANDARDS CORRELATIONS

The following tables show how each design challenge addresses Common Core State Mathematics Standards and International Technology and Engineering Standards. In the Common Core column, double asterisks (**) denote standards that are not expressly addressed by the design challenges, but that can be addressed by using optional suggestions included in the instructional text for that design challenge. References to the specific pages are included.

Common Core State Standards for Mathematics (Grades 6–8)[1]

Mathematical Practices

2. Reason abstractly and quantitatively.
   **See Optional CCSS Enhancement(s) on page 14.*
3. Construct viable arguments and critique the reasoning of others.
6. Attend to precision.

Standards

6.RP.1. Understand the concept of a ratio and use ratio language to describe a ratio relationship between two quantities.

6.RP.3. Use ratio and rate reasoning to solve real-world and mathematical problems, e.g., by reasoning about tables** of equivalent ratios, … or equations.**
   **See Optional CCSS Enhancement(s) on page 14.*
   a. Make tables of equivalent ratios relating quantities with whole-number measurements, find missing values in the tables…. Use tables to compare ratios.**
   **See Optional CCSS Enhancement(s) on page 14.*
   b. Solve unit rate problems including those involving … constant speed.
   d. Use ratio reasoning to convert measurement units; manipulate and transform units appropriately when multiplying or dividing quantities.

6.NS.1. Interpret and compute quotients of fractions, and solve word problems involving division of fractions by fractions….

7.G.1. Solve problems involving scale drawings of geometric figures, including computing actual lengths and areas** from a scale drawing….
   **See Optional CCSS Enhancement(s) on page 14.*

---

[1]Common Core State Standards. Copyright 2010. National Governor's Association Center for Best Practices and Council of Chief State School Officers. All rights reserved.

COMMON CORE AND ITEEA STANDARDS CORRELATIONS (CONTINUED)

Common Core State Standards for Mathematics (Grades 6–8)	ITEEA Standards for Technological Literacy (STL)[2]
Mathematical Practices 1. Make sense of problems and persevere in solving them. 2. Reason abstractly and quantitatively. 6. Attend to precision. **Standards** **6.RP.1.** Understand the concept of a ratio and use ratio language to describe a ratio relationship between two quantities. **6.RP.3.a.** Make tables of equivalent ratios relating quantities with whole-number measurements, find missing values in the tables**…. Use tables to compare ratios. **See Optional CCSS Enhancement(s) on page 31.* **6.RP.3.b.** Solve unit rate problems…. **6.RP.3.d.** Use ratio reasoning to convert measurement units; manipulate and transform units appropriately when multiplying or dividing quantities. **6.EE.9.** Use variables to represent two quantities in a real-world problem that change in relationship to one another; write an equation to express one quantity, thought of as the dependent variable, in terms of the other quantity, thought of as the independent variable. Analyze the relationship between the dependent and independent variables using … tables, and relate these to the equation.** **See Optional CCSS Enhancement(s) on page 31.* **7.RP.2.a.** Decide whether two quantities are in a proportional relationship, e.g., by testing for equivalent ratios in a table or graphing on a coordinate plane and observing whether the graph is a straight line through the origin. **7.RP.2.b.** Identify the constant of proportionality (unit rate) in tables, … equations,** diagrams, and verbal descriptions of proportional relationships. **See Optional CCSS Enhancement(s) on page 31.* **7.RP.2.c.** Represent proportional relationships by equations. **7.G.1.** Solve problems involving scale drawings of geometric figures, including computing actual lengths and areas from a scale drawing and reproducing a scale drawing at a different scale. **8.F.1.** Understand that a function is a rule that assigns to each input exactly one output. The graph of a function is the set of ordered pairs consisting of an input and the corresponding output.	**1F** New products and systems can be developed to solve problems or to help do things that could not be done without the help of technology. **1G** The development of technology is a human activity and is the result of individual and collective needs and the ability to be creative. **1H** Technology is closely linked to creativity, which has resulted in innovation. **2R** Requirements are the parameters placed on the development of a product or system. **2S** Trade-off is a decision process recognizing the need for careful compromises among competing factors. **8E** Design is a creative planning process that leads to useful products and systems. **8F** There is no perfect design. **8G** Requirements for design are made up of criteria and constraints. **9F** Design involves a set of steps, which can be performed in different sequences and repeated as needed. **9G** Brainstorming is a group problem-solving design process in which each person in the group presents his or her ideas in an open forum. **9H** Modeling, testing, evaluating, and modifying are used to transform ideas into practical solutions. **11H** Apply a design process to solve problems in and beyond the laboratory-classroom. **11J** Make two-dimensional and three-dimensional representations of the designed solution. **11K** Test and evaluate the design in relation to pre-established requirements, such as criteria and constraints, and refine as needed. **11L** Make a product or system and document the solution. **20F** The selection of designs for structures is based on factors such as building laws and codes, style, convenience, cost, climate, and function.

[2]International Technology and Engineering Educators Association (ITEEA). 2007. *Standards for Technological Literacy: Content for the Study of Technology.* (Third ed.) Virginia.

COMMON CORE AND ITEEA STANDARDS CORRELATIONS (CONTINUED)

Common Core State Standards for Mathematics (Grades 6–8)	ITEEA Standards for Technological Literacy (STL)
Mathematical Practices 1. Make sense of problems and persevere in solving them. 7. Look for and make use of structure. 8. Look for and express regularity in repeated reasoning. **Standards** **6.EE.9.** Use variables to represent two quantities in a real-world problem that change in relationship to one another; write an equation to express one quantity, thought of as the dependent variable, in terms of the other quantity, thought of as the independent variable. Analyze the relationship between the dependent and independent variables** using graphs and tables, and relate these to the equation. ***See Optional CCSS Enhancement(s) on page 65.* **6.G.1.** Find the area of right triangles, other triangles, special quadrilaterals, and polygons by composing into rectangles or decomposing into triangles and other shapes; apply these techniques in the context of solving real-world and mathematical problems. **6.G.2.** Find the volume of a right rectangular prism with fractional edge lengths by packing it with unit cubes of the appropriate unit fraction edge lengths**…. Apply the formulas $V = lwh$ and $V = bh$ to find volumes of right rectangular prisms with fractional edge lengths in the context of solving real-world and mathematical problems. ***See Optional CCSS Enhancement(s) on page 60.* **6.G.4.** Represent three-dimensional figures using nets made up of rectangles…, and use the nets to find the surface area of these figures. Apply these techniques in the context of solving real-world and mathematical problems.	**1F** New products and systems can be developed to solve problems or to help do things that could not be done without the help of technology. **1G** The development of technology is a human activity and is the result of individual and collective needs and the ability to be creative. **1H** Technology is closely linked to creativity, which has resulted in innovation. **2R** Requirements are the parameters placed on the development of a product or system. **2S** Trade-off is a decision process recognizing the need for careful compromises among competing factors. **8E** Design is a creative planning process that leads to useful products and systems. **8F** There is no perfect design. **8G** Requirements for design are made up of criteria and constraints.

	Common Core State Standards for Mathematics (Grades 6–8)	ITEEA Standards for Technological Literacy (STL)
DESIGN CHALLENGE 2: WE NEED WATER! *(Continued)*	**7.G.4.** Know the formulas for the area and circumference of a circle and use them to solve problems; give an informal derivation of the relationship between the circumference and area of a circle.** ****See Optional CCSS Enhancement(s) on page 62.* **7.G.6.** Solve real-world and mathematical problems involving area, volume, and surface area of two- and three-dimensional objects composed of triangles, quadrilaterals, polygons, cubes, and right prisms. **8.F.5.** Describe qualitatively the functional relationship between two quantities by analyzing a graph (e.g., where the function is increasing or decreasing, linear or nonlinear). Sketch a graph that exhibits the qualitative features of a function that has been described verbally.	**9F** Design involves a set of steps, which can be performed in different sequences and repeated as needed. **9G** Brainstorming is a group problem-solving design process in which each person in the group presents his or her ideas in an open forum. **9H** Modeling, testing, evaluating, and modifying are used to transform ideas into practical solutions. **11H** Apply a design process to solve problems in and beyond the laboratory-classroom. **11J** Make two-dimensional and three-dimensional representations of the designed solution. **11K** Test and evaluate the design in relation to pre-established requirements, such as criteria and constraints, and refine as needed. **11L** Make a product or system and document the solution.

DESIGN CHALLENGE 3: BALANCING ACT!

Common Core State Standards for Mathematics (Grades 6–8)	ITEEA Standards for Technological Literacy (STL)

Mathematical Practices
1. Make sense of problems and persevere in solving them.
5. Use appropriate tools strategically.
7. Look for and make use of structure.
 	
Standards
7.NS.1. Apply and extend previous understandings of addition and subtraction to add and subtract rational numbers; represent addition and subtraction on a horizontal or vertical number line diagram.**
***See Optional CCSS Enhancement(s) on page 99.*
  a. Describe situations in which opposite quantities combine to make 0.**
  ***See Optional CCSS Enhancement(s) on page 99.*
  b. Understand $p + q$ as the number located a distance $|q|$ from p, in the positive or negative direction depending on whether q is positive or negative.** Show that a number and its opposite have a sum of 0 (are additive inverses). Interpret sums of rational numbers by describing real-world contexts.
  ***See Optional CCSS Enhancement(s) on page 99.*
7.EE.4. Use variables** to represent quantities in a real-world or mathematical problem, and construct simple equations and inequalities to solve problems by reasoning about the quantities.
***See Optional CCSS Enhancement(s) on page 99.*
  a. Solve word problems leading to equations of the form $px + q = r$ and $p(x + q) = r$, where p, q, and r are specific rational numbers. Solve equations of these forms fluently.** Compare an algebraic solution to an arithmetic solution, identifying the sequence of the operations used in each approach.
  ***See Optional CCSS Enhancement(s) on page 99.*
  b. Solve word problems leading to inequalities of the form $px + q > r$ or $px + q < r$, where p, q, and r are specific rational numbers….**
  ***See Optional CCSS Enhancement(s) on page 99.*
	1F New products and systems can be developed to solve problems or to help do things that could not be done without the help of technology.
	
	1G The development of technology is a human activity and is the result of individual and collective needs and the ability to be creative.
	
	1H Technology is closely linked to creativity, which has resulted in innovation.
	
	2R Requirements are the parameters placed on the development of a product or system.
	
	2S Trade-off is a decision process recognizing the need for careful compromises among competing factors.
	
	8E Design is a creative planning process that leads to useful products and systems.
	
	8F There is no perfect design.
	
	8G Requirements for design are made up of criteria and constraints.
	
	9F Design involves a set of steps, which can be performed in different sequences and repeated as needed.
	
	9G Brainstorming is a group problem-solving design process in which each person in the group presents his or her ideas in an open forum.
	
	9H Modeling, testing, evaluating, and modifying are used to transform ideas into practical solutions.
	
	11H Apply a design process to solve problems in and beyond the laboratory-classroom.
	
	11J Make two-dimensional and three-dimensional representations of the designed solution.
	
	11K Test and evaluate the design in relation to pre-established requirements, such as criteria and constraints, and refine as needed.
	
	11L Make a product or system and document the solution.

PACING PLANNING GUIDE

*ESTIMATED TIME ASSUMES CLASS PERIODS OF 45–50 MINUTES.

Each design challenge takes 5 or more days to complete.

Section name	EDP step[1]	Page number	Estimated time*	Your time estimate
Shipwreck Survivors Team-Building Activity (optional)		1	Day 0	
Stranded! Prerequisite Math Skills (optional)		5	Day 0	
Stranded! Introduction		11	Day 1	
Where Are We?		13	Day 1	
A True Tale of Survival		16	Day 1	
Introducing the Engineering Design Process (EDP)		18	Day 1	
DESIGN CHALLENGE 1: A STORM IS APPROACHING!				
Define: Design criteria and constraints are defined.	1	26	Day 2	
Research: Inquiry activity on scaling three-dimensional objects; students determine and use scale to calculate amount of scale model materials.	2	28	Day 2	
Research: Students look at examples of structures.	2	35	Day 3, homework	
Brainstorm: Students individually sketch shelter structure design.	3	37	Day 3, homework	
Choose: Team decides on shelter structure design and draws design.	4	39	Day 3	
Build: Students build scale model of shelter.	5	41	Days 3–4	
Test: Students test scale model to see if criteria are met.	6	43	Day 4	
Communicate: Teams pair up to share designs and provide feedback and suggestions.	7	45	Days 4–5	
Redesign: Students answer questions about improving design.	8	47	Day 5	
DESIGN CHALLENGE 2: WE NEED WATER!				
Define: Design criteria and constraints are defined.	1	55	Day 1	
Research: Students calculate area of plane siding.	2	57	Day 1	
Research: Demonstration shows that same size and shape papers can be rolled up into cylinders with different volumes.	2	59	Day 1	
Research: Students measure the dimensions and volume of cylinders with bottoms, calculate the surface area, and graph the relationship between radius and volume and height and volume to find out which dimensions produce greatest volume.	2	62	Day 2	
Research: Similar investigation as previous research phase except using prisms; students compare/contrast with cylinder; also apply to hexagonal prism.	2	71	Day 3	

[1]To learn more about the engineering design process (EDP), see pages 125–126.

Section name	EDP step	Page number	Estimated time	Your time estimate
DESIGN CHALLENGE 2: WE NEED WATER! (*CONTINUED*)				
Brainstorm: Students individually sketch idea for water collector design.	3	76	Day 3, homework	
Choose: Team decides on water collector design and draws design.	4	78	Day 4	
Choose: Students calculate the surface area (and perhaps volume) of their team designs.	4	79	Day 4	
Build: Students build prototype of water collector.	5	82	Days 4–5	
Test: Students test water collector prototype to see if criteria are met.	6	84	Day 5	
Communicate: Teams pair up to share designs and provide feedback and suggestions.	7	86	Day 6	
Redesign: Students answer questions about improving design.	8	88	Day 6	
DESIGN CHALLENGE 3: BALANCING ACT!				
Define: Design criteria and constraints are defined.	1	96	Day 1	
Research: Students use the Math Balance seesaw to balance two objects and generalize the relationship of weight and distance of two balanced objects.	2	98	Day 1	
Research: Students use the Math Balance seesaw to balance more than two objects and generalize the relationship of weight and distance for any number of balanced objects.	2	101	Days 1–2	
Brainstorm: Individually balance various pairs of items/people represented by different numbers of tiles, and check using the Math Balance seesaw.	3	106	Day 2, homework	
Choose: Students design and record a step-by-step loading plan to put things/people on the canoe while keeping the canoe balanced at all times.	4	110	Day 3	
Build: Students organize stacks of tiles for loading.	5	112	Day 3	
Test: Students test loading plan to see if criteria are met.	6	114	Day 3	
Communicate: Students answer questions and share with class.	7	116	Day 4	
Redesign: After hearing from other teams, students answer questions to compare and improve design.	8	118	Day 4	

Teacher Page

SHIPWRECK SURVIVORS TEAM-BUILDING ACTIVITY *(OPTIONAL)*

Students work and communicate in teams during most of each design challenge. Some pilot teachers found it useful to do some team-building activities prior to the start of the unit. There is a different team-building activity in each of the *Building Math* books.

OBJECTIVES

- To compare the effectiveness of making decisions as an individual and as a team
- To practice communication skills
- To reflect on one's participation in a team setting

GROUP SIZE: 3 to 4 students

MATERIALS: a copy of the activity worksheet for each student

SETUP

- Arrange chairs and desks so students are seated in groups of 3 to 4 students.

PROCEDURES

1. Explain to the class that they will participate in a team-building activity that focuses on communication and decision making.
2. Distribute the activity worksheets to each student.
3. Read about the situation and challenge on the first page.
4. Read and answer any questions about the list of items on the second page.
5. Instruct students to individually choose and rank 10 items in order of priority to retrieve from the yacht. (5 minutes)
6. Arrange students in groups of 3 or 4 to discuss how they will choose and rank 10 items in order of priority as a team. (10 minutes) As you circulate around the room, identify natural "leaders" and "followers," and assess how well teams are communicating and making decisions. Use your observations to adjust groups as needed for *Stranded!* activities later.

7. Debrief the shipwreck activity with the following questions:

- Compare the list you made as an individual with the list you made as a team. How different are the two lists? Which list do you think is better? What are the advantages and disadvantages of making decisions as an individual compared to making decisions as a team?

- What was your role in this group? How much and how well did you contribute to the task? Was it hard or easy for you to communicate your ideas to the group? Did you feel that others were listening?

- How were decisions made? (vote? consensus? compromise?)

- Who influenced the decisions and how?

- How could better decisions have been made?

- How was conflict managed?

- How satisfied was each person with the decision? Ask each participant to rate his or her satisfaction out of 10.

- What have you learned about the functioning of this group? How well did the group work together?

- How would you do the activity differently if you were asked to do it again?

SHIPWRECK SURVIVORS TEAM-BUILDING ACTIVITY

You and your team are on a yacht, sailing in the Pacific Ocean. Suddenly, you get caught in a storm and crash on a coral reef! Jumping overboard, you all miraculously reach an island.

THE ISLAND

- 500 km away from a port
- small and uninhabited
- filled with tropical vegetation (coconuts, mangos, and bananas)
- rainy—almost every afternoon
- warm year round
- fish and shellfish seen just offshore at low tide

THE SITUATION

- All of you check your pockets and find two lighters, a pocketknife, and three watches.
- There's a lagoon between the beach and the coral reef. But because it's a long distance, you have to be a skilled swimmer to get across it. Only a few of you can make the trip.
- Equipment aboard the yacht (i.e., radio transmitter, radar, generators) was destroyed in the crash.
- The yacht will sink at high tide—8 hours from now.
- Together, you've made a list of the items you believe are on the boat.

THE TASK

- You all agree that the best swimmers must reach the wreck to get the items you need to survive before the yacht sinks.
- There are 30 items available from the yacht.
- The swimmers can only carry back 10 items.
- Make a list of priority items for the swimmers.

SHIPWRECK SURVIVORS TEAM-BUILDING ACTIVITY

RETRIEVABLE ITEMS

ax	hammer and nails	knives
fishing tackle	transistor radio	plastic buckets
bed sheets	blankets	shark repellant
canned food	beverages	first-aid kit
sunscreen	toilet articles	mirror
condensed milk	wooden planks	tool box
flippers and harpoon	water tanks	large plastic sheet
binoculars	life preservers	marine charts (maps)
bottles of rum	chocolate bars	rope
pistol and ammunition	cooking pots	mosquito netting
signal flares		

INDIVIDUAL RANKING

1. _____

2. _____

3. _____

4. _____

5. _____

6. _____

7. _____

8. _____

9. _____

10. _____

TEAM RANKING

1. _____

2. _____

3. _____

4. _____

5. _____

6. _____

7. _____

8. _____

9. _____

10. _____

Teacher Page

STRANDED! PREREQUISITE MATH SKILLS *(OPTIONAL)*

Students will be using the math skills listed below while doing the *Stranded!* activities. You may want to review these skills as a short warm-up exercise at the start of the class, or as a homework assignment before the skill is used. The skills in bold are reviewed in the activities on pages 7–9.

STUDENTS SHOULD BE ABLE TO				
MATH SKILL	WHERE ARE WE?	1. A STORM IS APPROACHING!	2. WE NEED WATER!	3. BALANCING ACT!
Interpret a scale on a map.	✓			
Use speed = distance/time to find one quantity given the other two quantities.	✓			
Measure length (in centimeters) using a ruler.		✓	✓	
Calculate the surface area of rectangular prisms and cylinders.		✓	✓	
Calculate the area of a triangle, a rectangle, a parallelogram, and a circle.		✓	✓	
Calculate the volume of rectangular prisms.		✓		
Use a given scale to determine scale model dimensions given actual dimensions.		✓		
Make a line graph.			✓	

* **Bold text indicates key objectives.**

Teacher Page *(continued)*

HOW TO USE THE WORKSHEETS

1. Speed = Distance/Time Activity (page 7)
   - The goal is to have students use the relationship *speed = distance/time* to find one of these quantities given the other two.
   - Walk through the two examples with students.
   - Assign students to do the problems on their own or with a partner.

2. Using a Scale Activity (pages 8–9)
   - The goal is to have students apply intuitive proportional reasoning by using drawings, and to maintain equal ratios by multiplying (or dividing) the numerator and denominator by the same number. Students should also be able to solve two-step proportion problems that call for unit conversion.
   - Walk through the two examples with students.
   - Assign students to do the problems on their own or with a partner.

SPEED = DISTANCE/TIME ACTIVITY

Read the examples below. Then use the relationship *speed = distance/time* to solve the word problems that follow.

EXAMPLE 1

Amy drove a car at 80 km per hour for 4 hours. How many miles did she travel during this time?

1. Identify the quantities given and the unknown quantity:

   80 km per hour is the speed.
   4 hours is the time.
   We are looking for the distance.

2. Rearrange the relationship so that distance is by itself on one side of the equation:

   $$\text{speed} = \frac{\text{distance}}{\text{time}}$$

   $$\text{speed} \bullet \text{time} = \frac{\text{distance}}{\text{time}} \bullet \text{time}$$

   $$\text{speed} \bullet \text{time} = \text{distance}$$

3. Replace the variables with the known quantities to find the unknown quantity:

   80 km/h $\bullet$ 4 hours = distance
   320 km = distance

EXAMPLE 2

How long did it take the yellow cab to travel 40 km at the average speed of 65 km per hour?

1. Identify the quantities given and the unknown quantity:

   65 km per hour is the speed.

   40 km is the distance.

   We are looking for the time.

2. Rearrange the relationship so that distance is by itself on one side of the equation:

   $$\text{speed} \bullet \text{time} = \text{distance}$$

   $$\frac{\text{speed} \bullet \text{time}}{\text{speed}} = \frac{\text{distance}}{\text{speed}}$$

   $$\text{time} = \frac{\text{distance}}{\text{speed}}$$

3. Replace the variables with the known quantities to find the unknown quantity:

   $$\text{time} = \frac{40 \text{ km}}{65 \text{ km/h}}$$

   $$\text{time} = \frac{8}{13} \text{ hour}$$

PROBLEMS

1. Jorge rides his bike at the rate of 20 kilometers per hour for 2.5 hours. How far did he go during this time?

2. Emma ran 10 kilometers in 1.5 hours. What was her average speed?

3. How long did it take Kim's boat to travel 20 km at the average speed of 60 km per hour?

USING A SCALE ACTIVITY

Read the examples below. Then use scale to determine scale model dimensions given actual dimensions in the problems that follow.

EXAMPLE 1

The scale of an actual building to its scale model is 10 meters : 2.4 cm. A door is 2 meters by 3 meters. What are its dimensions in the scale model?

10 meters																			

▼ ÷ into 20 parts to get 0.5 meters per part

0.5	0.5	0.5	0.5	0.5	0.5	0.5	0.5	0.5	0.5	0.5	0.5	0.5	0.5	0.5	0.5	0.5	0.5	0.5	0.5

2.4 cm																			

▼ ÷ into 20 parts to get 0.12 cm per part

0.12	0.12	0.12	0.12	0.12	0.12	0.12	0.12	0.12	0.12	0.12	0.12	0.12	0.12	0.12	0.12	0.12	0.12	0.12	0.12

Therefore, an actual length of 2 meters is represented by 0.48 cm in the scale model.

Divide 10 m by 5 to get 2 m. Also divide the denominator by 5 to keep the ratio (fraction) equivalent. Thus, 2.4 cm ÷ 5 is 0.48 cm. Therefore, 0.48 cm in the model represents 2 m in the actual building.

$$\frac{10 \text{ meters} \div 5}{2.4 \text{ cm} \div 5} = \frac{2 \text{ meters}}{0.48 \text{ cm}} \qquad \boxed{\frac{\text{actual measure}}{\text{scaled measure}}}$$

For the other dimension of the door, multiply 10 meters by $\frac{10}{3}$ to get 3 m. Therefore, multiply 2.4 cm by $\frac{10}{3}$ to keep the ratio equivalent, thus getting 8 cm, which represents 3 m of actual length.

$$\frac{10 \text{ meters} \times \frac{10}{3}}{2.4 \text{ cm} \times \frac{10}{3}} = \frac{3 \text{ meters}}{8 \text{ cm}}$$

EXAMPLE 2

The scale of a plane to its scale model is 1 m : 6 cm. If the wing is 45 meters long, how long is it in the scale model?

$$\frac{1 \text{ m} \cdot 45}{6 \text{ cm} \cdot 45} = \frac{45 \text{ m}}{270 \text{ cm}}$$

Since 1 m represents 6 cm, then 45 m would represent 270 cm, calculated from multiplying 45 and 6.

USING A SCALE ACTIVITY *(CONTINUED)*

PROBLEMS

1. The scale of a truck to its scale model is 7.5 m : 5 cm. If the height of the truck is 2.25 m, what is the height of the model truck?

2. The scale of a tower to its scale model is 1 meter : 5 centimeters. If the tower is 6 meters by 20 meters, what are the dimensions of the scale model?

3. The scale of a stone memorial to its scale model is 1 meter : 12 cm. If the memorial is in the shape of a cube with a volume of 1 cubic meter, what is its volume in the scale model? (*Hint:* What are the dimensions of the scale model of the cube?)

WRITING HEURISTICS, OR RULES OF THUMB

Heuristics are rules of thumb people follow in order to make judgments quickly and efficiently.

For each design challenge per class, keep a list on chart paper of research results and other suggestions to consider when making design decisions to meet criteria and constraints. The list should be revisited and added to or refined when students reflect on and discuss the results of their research. Encourage students to use the list when they brainstorm and choose a design. The list should help them find a design that would successfully meet the criteria and constraints.

EXAMPLE

Rules of Thumb for Design Challenge 2: We Need Water!

- To maximize the volume of a cylinder with a given surface area, make the radius equal to the height. You can get more volume making a cylinder than a prism if you have a certain amount of material.

- Fold up the edges of the base and secure with tape to keep the container from "leaking."

Teacher Page

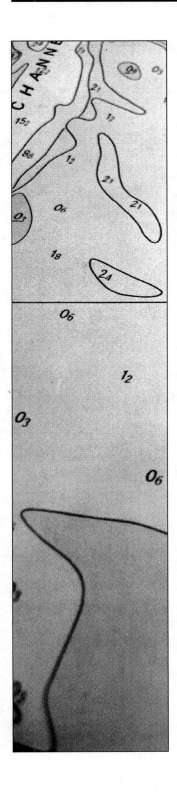

STRANDED! INTRODUCTION

OBJECTIVE: Students will read and understand the story line for the three design challenges in *Stranded!*

CLASS Tell the class to imagine that they are on an airplane over an ocean. Ask: "What would be the worst thing that could go wrong in that situation?" After allowing for a few responses, explain to students that if they were to be stranded on an island after a plane crash, they would find math and engineering skills very useful for survival. Over the next week, they will find out why. Read or ask a student to read aloud the Introduction on the next page.

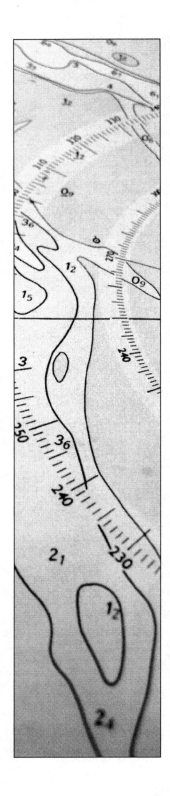

STRANDED! INTRODUCTION

Your school field trip to New Zealand to see where *The Lord of the Rings* movies were filmed has taken a turn for the worse! Your initial travel from your school to the San Francisco International Airport was fine. Your flight from San Francisco to New Zealand started off as scheduled. The plane was on course until a severe thunderstorm caused both the engine and the radio to fail. The plane was forced to crash somewhere in the South Pacific. The plane was torn to shreds, but the emergency raft remained intact. You and your classmates climbed into the raft and drifted on the ocean for several hours. You finally washed ashore on what seemed to be a deserted island.

- **WHERE ARE WE?**
  Using some clues, your math skills, and a map, figure out approximately where the deserted island is located.

- **INTRODUCING THE ENGINEERING DESIGN PROCESS (EDP)**
  What is engineering? What does an engineer do?

- **DESIGN CHALLENGE 1: A STORM IS APPROACHING!**
  Design a shelter to protect you and your team from a storm.

- **DESIGN CHALLENGE 2: WE NEED WATER!**
  Design a water collector with enough capacity for you and your team.

- **DESIGN CHALLENGE 3: BALANCING ACT!**
  Design a loading plan that can keep people and objects balanced in a canoe.

Teacher Page

WHERE ARE WE?

OBJECTIVES

Students will:

- interpret a scale on a map
- use proportional reasoning to calculate actual distance and drawn distance on a map according to a scale
- use the relationship speed = distance/time to find one quantity given the other two quantities
- solve a multistep problem
- use a ruler

MATERIALS

FOR THE CLASS

- map on transparency
- overhead projector, blank transparencies, and transparency marker

FOR EACH TEAM

- calculators
- rulers

1. **TEAMS** Explain to students that they will be working in teams to solve the first problem, which is to use clues to find the approximate location of the island where they are stranded. At the end of about 8 minutes, each team should be ready to show and explain where they drew the approximate location of the island on the map. As teams are working, circulate and note any differences in map drawings.

ASK EACH TEAM:

- Do you think that there is more than one correct answer?
- How can you represent all the possible correct answers?

2. **CLASS** Ask a team to present their answer by tracing their route on a map using an overhead projector and explaining any calculations they did to determine that route.

Teacher Page *(continued)*

ASK THE CLASS:
- Do you agree or disagree with the team's explanation?
- If you disagree, in what ways do you disagree and why?
- How did you come up with a way to represent all the possible places you could have drifted in the lifeboat?
- Did you come up with the same answer in a different way?

EXTENSION
Have students use an atlas to look up islands in this area.

OPTIONAL CCSS ENHANCEMENTS
To address additional aspects of the Common Core State Standards, direct students to construct data tables that represent time and distance. Instruct them to find the area within the circumference of the circle marking the island's potential location. Have students use letters and symbols to relate the problem challenge to formulas, i.e., $rt = d$, and the use of scale factors.

ASSESSMENT
Use the following problem to assess students' ability to solve a speed-distance-time problem. You might want to provide a simple map or have students draw their own visual representation of the walk.

Map scale: 2.5 cm = 0.5 km
Carol takes 20 minutes to walk to school at a speed of 0.15 km per hour.
   a. How far does Carol walk? (.05 km)
   b. Use the map scale to figure out the distance Carol walks as shown on the map. (0.25 cm)

WHERE ARE WE?

Where in the world is this island on which you have landed? Use the information below and the map on this page to estimate where the island is located. Show your work on a separate sheet of paper.

USEFUL INFORMATION

- The plane from San Francisco to New Zealand was traveling at a steady rate of 850 km/h before the storm hit.

- The storm hit the plane 9.5 hours into the flight.

- When the plane flew into the storm, the engine developed some problems. Fierce winds from the storm blew the plane approximately 800 km due south until it finally touched down in the ocean.

- On the emergency raft, you and your classmates drifted in the ocean, in an unknown direction, at a speed of approximately 15 km/h for one full day (24 hours) before reaching land.

Once you have figured out the approximate location of the island, draw a circle on the map to indicate this location.

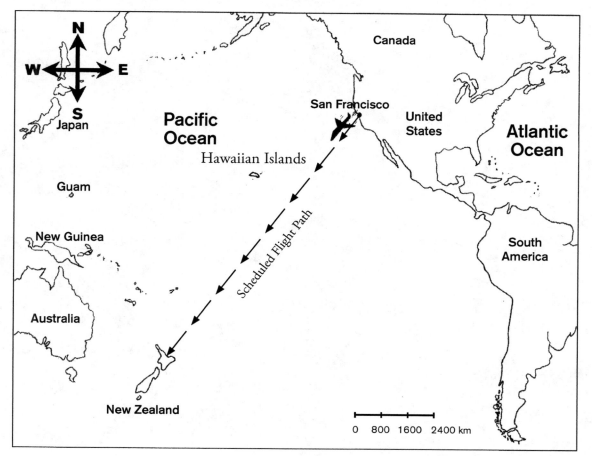

A TRUE TALE OF SURVIVAL

OBJECTIVE: Students will continue to read and understand the story line for the three design challenges in *Stranded!*

CLASS Read or ask a student to read the next page aloud.

ASK THE CLASS:

- What basic things do humans need to survive?
- What kinds of skills did Selkirk use to meet these basic needs?
- What math do you think is needed to use these skills?

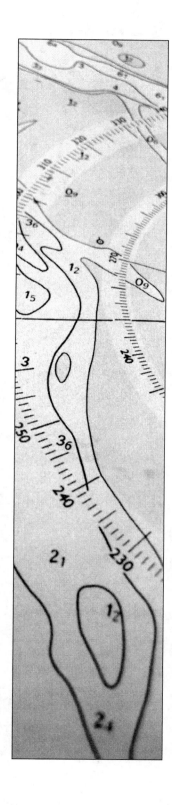

A TRUE TALE OF SURVIVAL

Before taking off, airplane pilots perform many checks to ensure a smooth flight. They check the weather over their route, the radio and equipment, and the overall condition of the airplane. Everything has to be working properly to ensure a safe flight. However, unexpected weather and equipment malfunctions can still surprise pilots and lead to disaster! There have been many stories of planes that, due to unexpected events, have crashed into the ocean, leaving their passengers stranded!

In 1704, Alexander Selkirk volunteered to be isolated on a deserted island. Selkirk remained stranded in the South Pacific for over four years before being rescued. When Selkirk first arrived on the deserted island, he only had a knife, a kettle, a gun, ammunition, and a book. However, he was able to make use of the island's resources to survive. Some might call him a natural-born engineer. Using pimento wood, he was able to create fire for cooking and light. The numerous wild animals he hunted provided food, as well as materials for the clothing that he designed and constructed to keep warm. Selkirk also designed and built a shelter using the island's natural resources. Long grass created the roofing, goat skin made the walls within the hut, and wood supported the overall structure.

Although Selkirk managed all the necessities of survival, he could not overcome the terrifying loneliness of years of solitude. He did not have the resources to design human companionship. After being rescued, Selkirk recounted his remarkable tale of survival. His story became the inspiration for English author Daniel Defoe's famous novel Robinson Crusoe. Based on Selkirk's real-life adventures, Defoe's novel tells the story of one man's faith, courage, and ability to survive on a deserted island against the brutal forces of nature.

INTRODUCING THE ENGINEERING DESIGN PROCESS (EDP)

OBJECTIVE: Students will identify and order the steps of the engineering design process (EDP).

ESTIMATED TIME: 20 minutes

MATERIALS

- 1 set of EDP cards per pair or team of 3 to 4 students

BEFORE YOU TEACH

- Make sets of EDP cards by copying the EDP card templates (pages 20–21) back to back onto card stock, cutting them out, shuffling them, and tying each set with a rubber band.
- Organize students into pairs or teams of 3 to 4 students.

PROCEDURES

1. To get a sense for what students know and think about engineering, ask: "What is an engineer? What does an engineer do?" Students can brainstorm and write their ideas in the space on page 22. If students are struggling to respond to these questions, ask them to list some things that are made by people—for example, houses, roads, cars, televisions, and phones. Explain that engineers have a part in the design and construction of all these things and many more.

2. Explain that all engineers use the engineering design process to help them solve problems in an organized way. Explain that students will use this design process to solve problems in this unit.

3. Distribute one set of EDP cards to each team. Instruct teams to distribute the cards evenly among themselves and take turns reading aloud each step's description. The students' task is to correctly order the steps. Set the time limit to 3 minutes. When debriefing the activity, post each team's steps on the board and compare the lists. Where do teams agree and disagree? Where there is disagreement, ask the teams to explain their rationale for their particular orders. When revealing the "correct" order on pages 125–126, emphasize that the EDP is meant to be a set of guidelines to solve engineering design challenges, but engineers may not always follow all the steps in the same order all the time.

4.   Ask the following questions to help students think more about these steps:

   • Why is the step "Communicate" part of the design process? How is it an important step?
     Possible answer(s): It's important for engineers to communicate their design to other people so they can receive critical feedback and suggestions to improve the design.

   • What do you think happens after the last step, "Redesign"?
     Possible answer(s): The engineer may go back to an earlier step—which could be as early in the process as "Identify the Problem"—depending on how well the prototype meets specifications. Once the design has gone through several cycles of the design process, it may then be produced on the full-scale level and constructed for real-world use.

5.   The EDP matching exercise on page 23 gives students an opportunity to identify the EDP step used in a specific instance of a design challenge.

6.   Wrap up the lesson by explaining that students will use these steps to solve three engineering challenges in *Stranded!* while learning and reinforcing their math skills and understanding. Point out that the octagon on the right-hand corner of each student page shows where students are in the EDP. The description of each EDP step is also on pages 125–126.

DEFINE the problem. What is the problem? What do I want to do? What have others already done? Decide upon a set of specifications (also called "criteria") that your solution should have.

Conduct **RESEARCH** on what can be done to solve the problem. What are the possible solutions? Use the Internet, go to the library, conduct investigations, and talk to experts to explore possible solutions.

BRAINSTORM ideas and be creative! Think about possible solutions in both two and three dimensions. Let your imagination run wild. Talk with your teacher and fellow classmates.

CHOOSE the best solution that meets all the requirements. Any diagrams or sketches will be helpful for later EDP steps. Make a list of all the materials the project will need.

Use your diagrams and list of materials as a guide to **BUILD** a model or prototype of your solution.

TEST and evaluate your prototype. How well does it work? Does it satisfy the engineering criteria?

COMMUNICATE with your fellow peers about your prototype. Why did you choose this design? Does it work as intended? If not, what could be fixed? What were the trade-offs in your design?

Based on information gathered in the testing and communication steps, **REDESIGN** your prototype. Keep in mind what you learned from others in the communication step. Improvements can always be made!

RESEARCH	DEFINE
CHOOSE	BRAINSTORM
TEST	BUILD
REDESIGN	COMMUNICATE

INTRODUCING THE ENGINEERING DESIGN PROCESS (EDP)

1. What is engineering? What does an engineer do? Brainstorm and list some of your ideas in the space below.

2. Your team will be given some cards, each naming and describing a step in the engineering design process (EDP). Engineers use the EDP to solve design challenges, just like you will as you go through *Stranded!* Your task is to put these steps in a logical order, from Step 1 to Step 8. Be prepared to explain your reasoning for the order you choose.

   STEP 1: _____

   STEP 2: _____

   STEP 3: _____

   STEP 4: _____

   STEP 5: _____

   STEP 6: _____

   STEP 7: _____

   STEP 8: _____

3. Imagine you are part of a team that builds sails and uses them in boat races. Match each sentence to the appropriate step in the engineering design process.

SENTENCE	ENGINEERING DESIGN PROCESS STEP
a. You talk with other sailors to find out how their sails are made.	
b. Your team spends Saturday making a new sail.	
c. After you win the race, you explain the design of the sail to your competitors.	
d. A race is coming up, and your boat needs a new sail. The team decides that the sail must be waterproof, affordable, and strong enough to handle powerful winds.	
e. After meeting to discuss the different designs, your team decides on one design. You find a marina that sells sail material that is strong, waterproof, and cheap.	
f. The sail works pretty well, but when strong gusts of wind blow, the seams rip. Your team resews the seams using stronger stitching.	
g. One week before the race, your team tests the new sail.	
h. Each person on your team sketches a sail design.	

Design Challenge 1
A Storm Is Approaching!

INTRODUCTION

OBJECTIVE: Students will read and understand the problem presented for the first design challenge.

CLASS Read or ask a student to read aloud the introduction.

ASK THE CLASS:
- Have you ever been caught outside in the rain or other bad weather without shelter or an umbrella?
- Why is it important for people to have shelter?

Explain that the challenge in this activity is to design and build a model of a shelter on the island where they have been stranded.

INTERESTING INFO
What to do and not do in a lightning storm:

a. Don't take refuge under a tree.

b. Don't huddle with others.

c. Don't sit on the ground.

d. Don't try to "read" the sky.

e. Do help a lightning victim. (It is safe.)

TRY IT! You can calculate the distance (in kilometers) of a storm by counting the number of seconds starting from when you first see lightning until when you hear thunder, and then dividing the number of seconds by three.

Design Challenge 1

A Storm Is Approaching!

INTRODUCTION

Dark clouds are heading your way, and the wind is beginning to pick up. There is a loud crack of thunder coming from the distance. It looks like severe weather will be here in just a few hours. How will you protect yourself during the storm?

Teacher Page

1. DEFINE THE PROBLEM: A STORM IS APPROACHING!

OBJECTIVE: Students will read and understand the criteria and constraints of the design challenge.

CLASS Together, read Define the Problem. Make sure that students understand the engineering criteria and constraints.

INTERESTING INFO

Engineers must always carefully plan their experiments and follow a structured design plan. The explosion of the Chernobyl nuclear power plant is an example of the importance of planning. In 1986, engineers were testing one of the turbines in the power plant. They did not have a structured plan for the experiment and ignored many safety precautions. The results were catastrophic because they did not follow the engineering criteria and design steps. The power plant had a meltdown and exploded. There were 31 casualties from the explosion. The surrounding area is currently an uninhabited nuclear wasteland.

DEFINE

1. DEFINE THE PROBLEM: A STORM IS APPROACHING!

The thunderstorm is quickly approaching, so you and your teammates will need to think fast. You need to design and construct a shelter that can withstand strong winds, keep out the rain, and hold all of your team members.

ENGINEERING CRITERIA	
STURDY	Given three heavy gusts of wind, the shelter must not move, tip over, or be damaged in any way.
WATER RESISTANT	Given three squirts of water to simulate rain, the inside of the shelter must remain completely dry.
SPACIOUS	Each member of your team must have at least 1 cubic meter (m³) of space.

ENGINEERING CONSTRAINTS	
ACTUAL MATERIALS ON THE ISLAND	MATERIALS FOR BUILDING THE SHELTER MODEL
logs (20 logs per team; 3 meters long each)	craft sticks (12 cm long)
strip of plane siding that washed ashore (1 piece 2.5 meters × 4 meters per team)	aluminum foil
tarp from the rescue raft (1 piece 3 meters × 5 meters per team)	wax paper
rope that washed ashore (6 meters of rope per team)	string
mud (1 bucket filled with 1 cubic meter (m³) of mud per team)	clay

Teacher Page

2. RESEARCH THE PROBLEM: A STORM IS APPROACHING!
RESEARCH PHASE 1: SCALE MODELING

OBJECTIVES

Students will:
- identify similar three-dimensional objects
- identify corresponding dimensions of similar objects
- use a ruler to measure three-dimensional objects
- calculate surface area and volume of rectangular prisms
- analyze a table of values for patterns
- generalize patterns using symbols

MATERIALS

For each team:
- calculator
- 1 pair of similar rectangular prisms
- 1 ruler (metric)

BEFORE YOU TEACH

Follow the directions in Appendix A (pages 142–157) to create nets of rectangular prisms. For each prism, mark the base and label with its letter. You may need to duplicate some of the pairs of similar prisms if you have more than four groups. Each group should receive one pair of similar prisms.

Box	Depth (cm)	Width (cm)	Height (cm)
A	3	7	11
B	5	5	8
C	3	4	5
D	2	2	2
E	6	14	22
F	15	15	24
G	12	16	20
H	10	10	10

Teacher Page *(continued)*

1. **CLASS** Show students a collection of boxes. Place the side marked "base" facedown on the table. Orient the box so that the width dimension is the longer side of the base (from students' perspective, left to right edge), and the depth dimension is the shorter side of the base (from students' perspective, front to back edge). In the case where the base is a square, the depth and width labels are interchangeable. The height is the vertical distance (top to bottom). These definitions of the dimensions (width, depth, and height) are based on Eric Weisstein's World of Mathematics (http://mathworld.wolfram.com). Together, read question 1 and ask each team to identify one set of boxes that are similar (same shape, different sizes, corresponding dimensions are proportional).

2. **TEAMS** Give students about 10 minutes to work in teams to complete questions 1 and 2.

3. **CLASS** First, make a table on the board like the one below using students' results from question 1. Second, use the table to ask students if they see a relationship between the scale factor and the area factor increase. In other words, given the scale factor, can they find a rule that will tell them how much larger the area will be? How much larger the volume will be? Or use the input-output game using the table below: If you put in 2, you get 4. If you put in 3, you get 9, and so forth. What's happening in between to change the input number to the output number? Repeat to find the cube rule for scaling volume.

Boxes pair	Scale factor (ratio of the corresponding sides of the large box to the small box)	How much larger is the area of the base of the large box than that of the small box?	How much larger is the volume of the large box than that of the small box?
A & E	2	4	8
B & F	3	9	27
C & G	4	16	64
D & H	5	25	125

4. **CLASS** Optional: Connect scale factors with graphing. The following graph shows how length changes with a scale factor of 3. The x-axis shows the actual length (independent variable) and the y-axis shows the scaled length (dependent variable). Here, a length of 3 units yields a scaled length of 1 unit. Make sure to show that this linear relationship does not hold with area and volume.

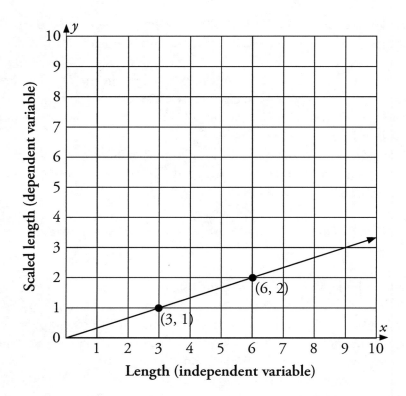

5. **CLASS** Optional: For additional practice with scale factor, have students write an equation using variables that they can use to predict scaled length given the length of an object and the scale factor. Copy or project the following example table on the board. Look at the table together and discuss.

Length	Scale factor	Scaled length
1 cm	3	3 cm
2 cm	3	6 cm
5 cm	3	15 cm
6 cm	3	18 cm
L	SF	SL

ASK THE CLASS:

- Do you see any patterns here? (The first column times the scale factor equals the third column.)
- Can we use this pattern to predict other scaled lengths? What if we start out with a length of 25 cm—what do we end up with after we scale it? (75 cm)
- Look at the last row of this table. It has letters instead of numbers. What do these letters stand for? (L = length; SF = scale factor; SL = scaled length)
- Can we use these letters to write an equation that will help us figure out a scaled length for any given length? What would that equation look like?

6.   **TEAMS**   Give students about 5 minutes to come up with the equation
($L \cdot SF = SL$).

7.   **TEAMS**   Draw or project the following table on the board. Have students use their
equation to fill in the values.

Length	Scale factor	Scaled length
1 cm	2	2 cm
2 cm	3	
2 cm	4	
5 cm	4	
1 cm	5	

OPTIONAL CCSS ENHANCEMENTS

To address additional aspects of the Common Core State Standards, supply additional
missing value problems, and/or direct students to write equations that express their
findings about scale factors.

Name
STUDENT PAGE

2. RESEARCH THE PROBLEM: A STORM IS APPROACHING!
RESEARCH PHASE 1: SCALE MODELING

In order to build an accurate scale model of your shelter, you will begin the research phase by investigating the effect of scaling on one, two, and three dimensions.

1.   You have been given a number of boxes. Examine each box, and then identify two boxes that are different in size, but seem to have the same shape and proportions. The smaller box should look like a miniature version of the larger.

   a.   Before you measure the dimensions of the box, identify corresponding widths, depths, and heights of the two boxes. How many times larger is the height of the large box compared to the small box? _____

	SMALL BOX	**LARGE BOX**
WIDTH (CM)		
DEPTH (CM)		
HEIGHT (CM)		

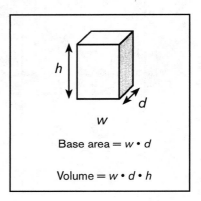

Base area $= w \cdot d$

Volume $= w \cdot d \cdot h$

   b.   Calculate the area of the base of the small box: _____ cm
   Calculate the area of the base of the large box: _____ cm
   How many times larger is the area of the base of the large box compared to the small box? _____

   c.   Calculate the volume of the small box: _____ cm
   Calculate the volume of the large box: _____ cm
   How many times larger is the volume of the large box compared to the small box? _____

2.   Based on your calculations above, what conclusions can you make about scaling one-dimensional, two-dimensional, and three-dimensional objects?

_____

_____

2. RESEARCH THE PROBLEM: A STORM IS APPROACHING!
RESEARCH PHASE 1: SCALE MODELING

OBJECTIVE: Students will use a scale to calculate the amount of materials available for building a scale model.

1. [**CLASS**] Read question 3 together. Have students review the list of actual materials (logs, plane siding, etc.) and point out the dimensions. Demonstrate using a meter stick or tape measure to show how long each log would actually be.

   ASK THE CLASS:
   - Why do engineers build scale models to test their designs before building the real thing?
     Possible answer(s):
     a. A smaller model means less quantity of materials—allowing engineers to try out several designs.
     b. Using a scale keeps the dimensions proportional so that the model looks exactly like a miniature version of the real thing.
   - If the real size of the log is 3 meters and the craft stick you are using to represent the log in the scale model is 12 centemeters, what scale could you use?
     Possible answer(s): If 3 meters is scaled down to 12 centimeters, then 1 meter can be represented by 4 centimeters. The scale would be 4 cm = 1 m.

2. [**TEAMS**] Give teams about 10 minutes to answer questions 3 and 4 and calculate the scale model dimensions using the scale they selected.

   - 3c: Students might incorrectly reason that since 4 cm : 1 m, then 4 cm^3 : 1m^3. Remind students about what they found out earlier about scaling three dimensions: The result is the scale factor cubed. Thus, 1 m^3 is scaled to 4^3 cm^3 or 64 cm^3. Another way to get the correct answer is to convert the individual lengths (1 m • 1 m • 1 m becomes 4 cm • 4 cm • 4 cm), and then multiply the lengths together to find the volume.

3. [**CLASS**] If students need additional practice using a scale, see the activities on pages 8–9.

2. RESEARCH THE PROBLEM: A STORM IS APPROACHING!

RESEARCH PHASE 1: SCALE MODELING

3.   You will be building a scale model of the shelter you design. Therefore, you must determine a scale for your design. As a group, decide on a scale for your model, and then answer the questions below.

   a. What scale did you choose? _____

   b. Why did your group choose this scale?

   c. Your shelter must provide at least 1 cubic meter (1 meter • 1 meter • 1 meter) of personal space for each team member. Based on your scale, how much personal space must be available for each person in the model shelter, in cubic centimeters? Show your work.

4.   Using your scale and the actual amounts of available material, figure out how much of each material you need for your model.

ACTUAL MATERIALS ON THE ISLAND	AMOUNT OF EACH MATERIAL FOR YOUR MODEL
logs (20 logs per team; 3 meters long each)	a. How many 12-cm craft sticks should you get? _____ sticks
strip of plane siding that washed ashore (1 piece 2.5 meters × 4 meters)	b. What size piece of aluminum foil do you need? _____ cm × _____ cm
tarp from the rescue raft (1 piece 3 meters × 5 meters)	c. What size piece of wax paper do you need? _____ cm × _____ cm
rope that washed ashore (6 meters of rope per team)	d. How much string do you need? _____ cm
mud (1 bucket filled with 1 cubic meter (cm³) of mud per team)	e. How much clay do you need? _____ cubic cm (cm³)

2. RESEARCH THE PROBLEM: A STORM IS APPROACHING!
RESEARCH PHASE 2: SHELTERS

OBJECTIVE: Students will look at drawings of different shelters and discuss their advantages and disadvantages as they consider how they will use their materials to design a shelter.

TEAMS Give teams time to discuss what kind of structure they want to build using the pictures on page 36. Students can also design their own.

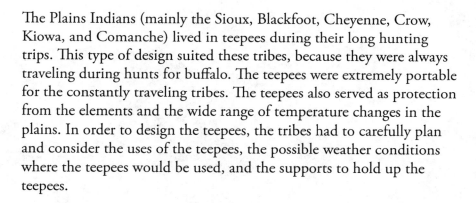

INTERESTING INFO

The choice of materials for the construction of a building is extremely important. The designer must consider the location of the building, the uses of the building, the possible weather conditions, and the design of the supports.

The Plains Indians (mainly the Sioux, Blackfoot, Cheyenne, Crow, Kiowa, and Comanche) lived in teepees during their long hunting trips. This type of design suited these tribes, because they were always traveling during hunts for buffalo. The teepees were extremely portable for the constantly traveling tribes. The teepees also served as protection from the elements and the wide range of temperature changes in the plains. In order to design the teepees, the tribes had to carefully plan and consider the uses of the teepees, the possible weather conditions where the teepees would be used, and the supports to hold up the teepees.

2. RESEARCH THE PROBLEM: A STORM IS APPROACHING!
RESEARCH PHASE 2: SHELTERS

You and your teammates are in luck! A few of the magazines that were on the plane have washed up on shore. Most of the pages have been ruined, but a page of *Survival Digest*, containing pictures of different types of shelters you can build while camping, is still intact. Use the pictures to help you decide what type of structure might be best for your team's shelter on the island.

3. BRAINSTORM POSSIBLE SOLUTIONS: A STORM IS APPROACHING!

OBJECTIVE: Students will individually brainstorm and sketch a shelter design.

INDIVIDUALS Instruct students to individually complete numbers 1 and 2 after reviewing the criteria and constraints.

INTERESTING INFO

During ocean cruises, accidents may happen, and passengers may have to abandon the ship. In such a case, life rafts are needed for all passengers and crewmembers. A basic life raft will be able to stay afloat, and more advanced ones may have sails or oars. The minimum amount of space that the U.S. Coast Guard requires per person in the life raft is about 1 cubic meter. Once inflated, life rafts usually last about four to five days. They also need to be replaced every fifteen years.

3. BRAINSTORM POSSIBLE SOLUTIONS: A STORM IS APPROACHING!

INDIVIDUAL DESIGN

1. In step 2, you decided on a scale for your model and determined how much material you would need for your shelter scale model. Please list these amounts in the space below.

ENGINEERING CONSTRAINTS

- 12-cm-long craft sticks: _____
- aluminum foil: _____ cm • _____ cm
- wax paper: _____ cm • _____ cm
- string: _____ cm
- clay: _____ cm³

2. Sketch one possible shelter design in the space below. Figure out what dimensions (height, length, width) would work given the available amount of material. Please label all parts. Be creative!

ENGINEERING CRITERIA

STURDY
→ Given three heavy gusts of wind, the shelter must not move, tip over, or be damaged in any way.

WATER-RESISTANT
→ Given three squirts of water to simulate rain, the inside of the shelter must remain completely dry.

SPACIOUS
→ Each member of your team must have at least 1 cubic meter (m³) of space.

4. CHOOSE THE BEST SOLUTION: A STORM IS APPROACHING!

OBJECTIVE: Students will share their individual brainstorm designs and decide on a team design.

MATERIALS

For each team:
- craft sticks
- aluminum foil
- wax paper
- string
- clay

1. **TEAMS** Team members should share their brainstorm designs and decide on a team design. Give each team a set of sample material to experiment with as they decide on their design. They will receive a final set of material when they get to the building prototype phase.

> **CLASSROOM MANAGEMENT TIPS**
> - If students need more structure to share their individual shelter designs from the Brainstorm step, go over a list of things for each person to share (for example, position and function of materials, shape of structure and rationale, how materials are joined, etc.); a set time for each person to share (1 minute); and behavior expectations for teammates during sharing (for example, no interruptions, no comments until everyone has shared, active listening).
> - Discuss with students how to come to an agreement on a team design. Team members can take turns discussing pros and cons of each design, identify commonalities in designs, compromise on areas of disagreement, and either vote or try to reach consensus.

2. **TEAMS** After each team finishes their design, check to make sure that their drawing is clearly labeled with names and dimensions of materials. If their drawing is complete, then they may begin building their prototype. If not, they need to continue working on the design until it is complete.

> **ASSESSMENT**
> Introduce students to the Rubric for Engineering Drawings on page 128 by showing students examples of student work on pages 134–141, and using the rubric to grade each drawing. You can first assess one drawing with the whole class using the rubric and then have students work in pairs to assess the other drawings. Debrief as a whole class. Students can use the rubric to self-assess their own drawings of the shelter design.

4. CHOOSE THE BEST SOLUTION: A STORM IS APPROACHING!

TEAM DESIGN

Discuss your ideas as a group. Decide on a team design that meets the design criteria and constraints, and sketch your design in the space below. Please label the dimensions (length, width, height) of your model, the materials you are using, and the scale you are using on your diagram. Use the Rubric for Engineering Drawings provided by your teacher to check the quality and completeness of your drawing. Practice using the rubric drawings on provided by your teacher.

5. BUILD A PROTOTYPE/MODEL: A STORM IS APPROACHING!

OBJECTIVE: Students will build their teams' shelter designs.

MATERIALS
For each team:
- calculator
- craft sticks
- aluminum foil
- wax paper
- string
- clay
- ruler (metric)

TEAMS Take away the sample materials and give teams a final set of materials according to the quantity calculated using each team's scale. As teams begin building their prototype, instruct them to write down or draw any changes for the original design.

ASSESSMENT
Use the Rubric for Prototype/Model on page 131 to assess completeness and craftsmanship of students' models.

TIP
Warn teams that they only have the materials they are given. If students have problems for whatever reason (for example, they run out of materials, something gets torn, or materials are measured incorrectly), they cannot get replacement materials, just as if they were stranded in real life. They will have to make do with whatever materials they have left. Make sure they record any changes to their design.

5. BUILD A PROTOTYPE/MODEL: A STORM IS APPROACHING!

Collect all necessary materials, and build your model according to the design you chose in step 4. As you build your model, write down any changes to your original design. Use the Rubric for Prototype/Model provided by your teacher to assess your work.

Teacher Page

6. TEST YOUR SOLUTION: A STORM IS APPROACHING!

OBJECTIVE: Students will follow testing procedures to test their shelter prototypes.

MATERIALS
For each team (taking turns):
- spray bottle with water
- stiff cardboard at least 21 cm × 26 cm

CLASS Instruct each team to test their shelter design. Use the stiff cardboard to fan the structure three times to simulate wind gusts. If there is still time after all the prototypes have been tested, teams can evaluate their designs by answering questions 1 and 2.

ASSESSMENT
Use the Rubric for Test, Communicate, and Redesign Steps on page 132 to assess how well students followed testing procedures.

INTERESTING INFO
Hurricanes are severe tropical storms in the Atlantic Ocean, Caribbean Sea, Gulf of Mexico, and the eastern Pacific Ocean. Hurricanes have a minimum speed of 119 kilometers per hour. The following shows the classification for hurricanes:
- Category One: winds 119 to 153 km per hour
- Category Two: winds 154 to 177 km per hour
- Category Three: winds 178 to 209 km per hour
- Category Four: winds 210 to 249 km per hour
- Category Five: winds greater than 249 km per hour

TEST

6. TEST YOUR SOLUTION: A STORM IS APPROACHING!

Perform the following tests on your model. For each test, check "yes" if your shelter passes the test and "no" if it fails. You may use the Rubric for Test, Communicate, and Redesign Steps provided by your teacher to assess your work on the next few pages.

STURDINESS TEST

Your teacher will create three gusts of wind near your shelter.

Did the shelter stay in one place and remain intact? ☐ yes ☐ no

SPACIOUSNESS TEST

- Each team member should have at least 1 cubic meter of personal space.
- What is the scale factor for your model shelter? _____
- Based on the scale factor, how much personal space should each person have in your model shelter? _____
- How much space does your shelter have? _____
- Alternatively, you can use a paper cube scaled to represent 1 cubic meter of personal space for each person on your team. See if these paper cubes fit inside your shelter.

Is there enough room in your shelter for each person to have this much personal space? ☐ yes ☐ no

WATER-RESISTANCE TEST

Place pieces of paper or scaled paper cubes that represent each team member inside your shelter. Your teacher will spray your shelter with three squirts of water to simulate rain.

Did the paper stay completely dry? ☐ yes ☐ no

EVALUATE YOUR DESIGN

1. Based on the three tests, how well did your shelter meet the design criteria?

   _____

   _____

   _____

2. What are some areas where you can improve your shelter design?

   _____

   _____

   _____

© Museum of Science (Boston), Wong, Brizuela

7. COMMUNICATE YOUR SOLUTION: A STORM IS APPROACHING!

OBJECTIVE: Students will answer questions to reflect on their designs, present their designs to another team, and give feedback on the other team's design.

TEAMS Pair teams and instruct them to choose one person to describe their design to the other team and their evaluation from the testing page. Each team member is to offer at least one suggestion to the other team to improve the other team's design. Set the time limit at about 15 minutes.

COMMUNICATE

7. COMMUNICATE YOUR SOLUTION: A STORM IS APPROACHING!

Teams will now pair up to share their designs, and provide feedback and suggestions.

Nominate one person in your group to describe your model and share your evaluation (from step 6) with the other team.

Each member of the other team should provide one suggestion for how to improve your design. Record these suggestions in the space below.

Suggestion 1:

_____

_____

_____

Suggestion 2:

_____

_____

_____

Suggestion 3:

_____

_____

_____

Suggestion 4:

_____

_____

_____

8. REDESIGN AS NEEDED: A STORM IS APPROACHING!

OBJECTIVE: Students will answer questions to consider how they can redesign their shelters.

1. **TEAMS** Instruct teams to use the suggestions they received from the other team to improve their original design by describing the changes. Set the time limit to about 7 to 8 minutes.

2. **CLASS** *Optional:* If there is time, invite each team to give a short persuasive presentation on why their shelter design should be chosen. Teams should explain how their design best meets the criteria and constraints, as well as highlight other unique and useful features of their design.

3. **CLASS** Have students vote on which shelter they would most like to have on the island.

4. **CLASS** Ask the questions below to wrap up this activity.

ASK THE CLASS:
- How did you use scale in this activity?
  Possible answer(s): We used scale to calculate how much material we needed to build the scale model of a shelter.

- What other math skills did you use to build the prototype?
  Possible answer(s): We used the ruler to measure the materials we used to build the scale model. We also calculated the amount of materials using basic operations and area formulas of different shapes as we drew our design and decided on dimensions. We also calculated the volume to make sure that there is enough space inside to fit all the people on our team.

8. REDESIGN AS NEEDED: A STORM IS APPROACHING!

As a group, review the other team's suggestions.

1. Based on their suggestions, how can you change or improve your design?

    _____

    _____

    _____

    _____

    _____

    _____

    _____

2. The class will vote on which team's shelter they would most like to have on the island. Before the vote, describe the necessary improvements and prepare a persuasive argument about why your team's design should be chosen.

    _____

    _____

    _____

    _____

    _____

    _____

    _____

3. Consider the shelter that was voted for by the class. Why do you think this shelter was chosen over the others?

    _____

    _____

    _____

    _____

    _____

    _____

INDIVIDUAL SELF-ASSESSMENT RUBRIC: A STORM IS APPROACHING!

OBJECTIVE: Students will use a rubric to individually assess their involvement and work in this design challenge.

> **ASSESSMENT**
> Assign this reflection exercise as homework. You can write your comments on the lines below the self-assessment, and/or use this in conjunction with the student participation rubric on page 133.

Name
STUDENT PAGE

INDIVIDUAL SELF-ASSESSMENT RUBRIC: A STORM IS APPROACHING!

Use this rubric to reflect on how well you met behavior and work expectations during this activity. Check the box next to each expectation that you successfully met.

LEVEL 1	LEVEL 2	LEVEL 3	LEVEL 4	BONUS POINTS
Beginning to meet expectations	Meets some expectations	Meets expectations	Exceeds expectations	
☐ I was willing to work in a group setting.	☐ I met all of the Level 1 requirements.	☐ I met all of the Level 2 requirements.	☐ I met all of the Level 3 requirements.	☐ I helped resolve conflicts on my team.
☐ I was respectful and friendly to my teammates.	☐ I recorded the most essential comments from other group members.	☐ I made sure that my team was on track and doing the tasks for each activity.	☐ I helped my teammates understand the things that they did not understand.	☐ I responded well to criticism.
☐ I listened to my teammates and let them fully voice their opinions.	☐ I read all instructions.	☐ I listened to what my teammates had to say and asked for their opinions throughout the activity.	☐ I was always focused and on task: I didn't need to be reminded to do things; I knew what to do next.	☐ I encouraged everyone on my team to participate.
☐ I made sure we had the materials we needed and knew the tasks that needed to be done.	☐ I wrote down everything that was required for the activity.	☐ I actively gave feedback (by speaking and/or writing) to my team and other teams.	☐ I was able to explain to the class what we learned and did in the activity.	☐ I encouraged my team to persevere when my teammates faced difficulties and wanted to give up.
	☐ I listened to instructions in class and was able to stay on track.	☐ I completed all the assigned homework.		☐ I took advice and recommendations from the teacher about improving team performance and used feedback in team activities.
	☐ I asked questions when I didn't understand something.	☐ I was able to work on my own when the teacher couldn't help me right away.		☐ I worked with my team outside of the classroom to ensure that we could work well in the classroom.
		☐ I completed all the specified tasks for the activity.		

Approximate your level based on the number of checked boxes: _____   Bonus points: _____

Teacher comments: _____

Teacher Page

TEAM EVALUATION: A STORM IS APPROACHING!

OBJECTIVE: Students will evaluate and discuss how well they worked in teams.

ASSESSMENT

1. **INDIVIDUALS** Assign this reflection exercise as homework or during quiet classroom time.

2. **TEAMS** Instruct students to share their team evaluation reflections with one another and discuss how they can improve their teamwork during the next activity.

3. **CLASS** Point out any good examples of teamwork and areas to improve during the next activity.

TEAM EVALUATION: A STORM IS APPROACHING!

How well did your team work together to complete the design challenge? Reflect on your teamwork experience by completing this evaluation and sharing your thoughts with your team. Celebrate your successes and discuss how you can improve your teamwork during the next design challenge.

RATE YOUR TEAMWORK. On a scale of 0–3, how well did your team do? 3 is **excellent;** 0 is **very poor.** Explain how you came up with that rating.

LIST THINGS THAT WORKED WELL. Example: We got to our tasks right away and stayed on track.

LIST THINGS THAT DID NOT WORK WELL. Example: We argued a lot and did not come to a decision that everyone could agree on.

HOW CAN YOU IMPROVE TEAMWORK? Make the action steps concrete. For example: We need to learn how to make decisions better. Therefore, I will listen and respond without raising my voice.

Teacher Page

Design Challenge 2
We Need Water!

INTRODUCTION

OBJECTIVE: Students will read and understand the problem presented for the second design challenge.

> **CLASS**

ASK THE CLASS:
- In the last activity, we took care of one of the basic human needs: shelter. What other basic human needs are there?
- How long can a human being survive without water?
- Where can we get fresh water?

Design Challenge 2

We Need Water!

INTRODUCTION

The average person cannot survive for more than one week without fresh water. The island is surrounded by salty ocean water, but there is no fresh water on the island. How will you survive?

1. DEFINE THE PROBLEM: WE NEED WATER!

OBJECTIVE: Students will read and understand the criteria and constraints of the design challenge.

CLASS Read the engineering criteria and constraints as a class and make sure students understand them.

INTERESTING INFO: Math problem

You want to build a rectangular fence to enclose your front yard. You only have 15 meters of fence, and the area that you want to enclose is 75 square meters. The engineering criteria is to create a fence enclosure with an area of 75 square meters, and the constraint is that you only have 15 meters of fence. It is mathematically possible to meet both the criteria and constraint?

ANSWER: No. The rectangle that maximizes area with the smallest perimeter is a square. Thus, the length of a side (w) of a square with area 75 square meters would be $\sqrt{75}$ or $5\sqrt{3}$, which is approximately 8.7 meters. Such a square would have a perimeter of $20\sqrt{3}$, or approximately 34.6 meters, thus exceeding the constraint of 15 meters.

1. DEFINE THE PROBLEM: WE NEED WATER!

It has been raining almost every day. Is there a way you could collect the rainwater for drinking? A large, jagged piece of plane siding has washed ashore. Could you use this material to create a water collection tool?

ENGINEERING CRITERIA

STRONG

→ It remains intact when filled to capacity.

LEAKPROOF

→ When filled, it does not leak. This means that it should have a bottom.

LARGE CAPACITY

→ It can hold at least 1,000 mL of water for each member of your team, and the more the better!

FREESTANDING

→ It can stand upright without anyone holding it.

ENGINEERING CONSTRAINTS

- scrap of plane siding (approximate measurements provided below)
- roll of tape (washed ashore)

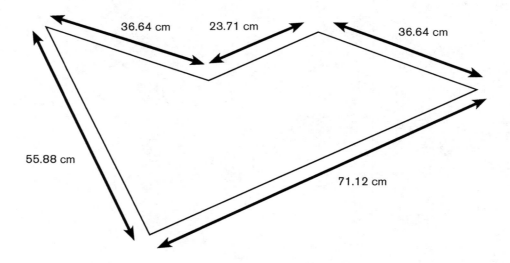

© Museum of Science (Boston), Wong, Brizuela

Teacher Page

2. RESEARCH THE PROBLEM: WE NEED WATER!
RESEARCH PHASE 1: FIND THE AREA OF THE PLANE SIDING

OBJECTIVE: Students will find the area of an irregular two-dimensional shape using strategies for finding the areas of triangles, rectangles, and parallelograms.

MATERIALS

For each team:

- a poster board in the exact shape and dimensions as the diagram on page 56

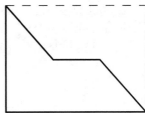

For each class:

- a sheet of chart paper for the Rules of Thumb list (see page 10)

BEFORE YOU TEACH

Draw the plane siding shape (exact dimensions) on poster board. Hint: Use poster board that is 55.88 cm by 71.12 cm so that two plane siding shapes make up the entire rectangle. Cut one out and use it to trace on additional poster boards. Each team gets one piece.

1. **TEAMS** Give each team a piece of poster board that is exactly the size and shape of the plane siding. Give teams 10 minutes to calculate the area. If a team finishes early, ask them to find another way to solve the problem. As teams are working, circulate to find two different methods of finding the area and ask those teams to share with the whole class later.

2. **CLASS** Invite two teams to share how they found the area of the poster board. Ask the rest of the class if they agree or disagree with the solution and why.

3. **CLASS** Introduce Heuristics or Rules of Thumb (see page 10). Ask students what they learned from the activity that they can add to the Rules of Thumb list.

Area of a triangle $= 0.5 \cdot b \cdot h$	
Area of square $= s^2$	
Area of a rectangle $= l \cdot w$	
Area of a trapezoid $= 0.5 \cdot h \cdot (b_1 + b_2)$	
Area of parallelogram $= b \cdot h$	

2. RESEARCH THE PROBLEM: WE NEED WATER!
RESEARCH PHASE 1: FIND THE AREA OF THE PLANE SIDING

Your group has been given the scrap of plane siding that you will use to make your water collector. It's important to know how much material you have to work with. What is the total area of this material in square centimeters (cm²)? Use the measurements of the poster board that represents the plane siding, NOT the measurements of the scale model drawing below.

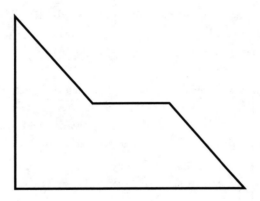

scale model of plane siding

Teacher Page

2. RESEARCH THE PROBLEM: WE NEED WATER!
RESEARCH PHASE 2: DOES SAME SURFACE AREA MEAN SAME VOLUME?

OBJECTIVE: Students will make conjectures and write about their reasoning on whether objects with the same surface area also have the same volume.

MATERIALS

For the class:
- 1 sheet of transparency cut in half so each piece 14.0 cm × 10.8 cm
- large flat box
- dry navy beans (at least 5 liters)
- tape

1. **CLASS/INDIVIDUALS** As a class, read question 1. Instruct students to individually think and write down an answer. Then give students 2 minutes to discuss their answers with a partner. Finally, ask a few individuals to share with the class what they were thinking and discussing. Many students will incorrectly think that if two containers are made with the same surface area, they should hold the same amount of water (have the same volume).

2. **CLASS** Show the class the two half-sheets of transparency and ask:
   - Are you convinced that these two sheets are the same size and shape (or congruent)?
   - What can you do to show that they are congruent?

   Possible answer(s): Put the two sheets on top of each other to show that there are no overlaps. Take one of the half-sheets of transparency, and join the top and bottom edges to form a "baseless" cylinder. The edges should meet exactly, with no gaps or overlap. With the other half-sheet of transparency the same size and aligned the same way, join the left and right edges to make another cylinder. Note that the two sheets have the same surface area. Stand both cylinders on a table. One of the cylinders will be tall and narrow; the other will be short and wide. Refer to the tall cylinder as cylinder X and the short one as cylinder Y. Mark each cylinder now to avoid confusion later.

ASK THE CLASS:
- "Do you think the two cylinders will hold the same amount? Or will one hold more than the other? If you think that one will hold more, which one will that be, and why?" Have students discuss in pairs or teams, and then share their predictions with the class, with explanations.

Teacher Page *(continued)*

Place cylinder X in a large flat box (the lid of a file box works well). Fill cylinder X to the top with beans. Place cylinder Y over cylinder X.

ASK THE CLASS:

- For those of you who predicted that the two cylinders will hold the same amount, what should happen if I lift cylinder X and all the beans in cylinder X go into cylinder Y?

  Possible answer(s): The beans would fill cylinder Y to the top.

Slowly lift cylinder X so that the beans empty into cylinder Y.

ASK THE CLASS:

- What does this show about which cylinder holds more? How can you tell?

  Possible answer(s): Since cylinder Y is only partway full, cylinder Y holds more than cylinder X.

ASK THE CLASS:

- "Was your prediction correct? Do the two cylinders hold the same amount? Why or why not? Can we explain why they don't?" These questions should spark an interesting discussion about the two cylinders.

  Possible answer(s): (At this point in the research phase, it is not crucial for students to be able to mathematically explain why. However, you can challenge advanced students to figure out why this works.) Volume (of a cylinder) = $\pi \cdot$ radius$^2 \cdot$ height. Since the radius is squared (multiplied by itself) and the height is only multiplied once, the radius has a greater effect on the volume. The radius is related to volume quadratically, whereas the height is only related to the volume linearly. Thus, cylinder Y's greater radius has a greater impact on the volume than cylinder X's greater height.

ASSESSMENT

INDIVIDUALS After the discussion, ask students to answer question 2 on their own. As they are working, you can circulate around the room to assess whether students understood the explanation and can write it in their own words.

OPTIONAL CCSS ENHANCEMENT

To address additional aspects of the Common Core State Standards, direct students to physically pack a prism with unit cubes.

---

2. RESEARCH THE PROBLEM: WE NEED WATER!
RESEARCH PHASE 2: DOES SAME SURFACE AREA MEAN SAME VOLUME?

1. Every group has the same plane siding. If each group uses all of the paper they have been given to make their water collector (without any overlapping pieces), will all the containers hold the same amount of water? Explain.

    _____

    _____

    _____

    _____

    STOP   **Wait for your teacher to tell you to continue.**

2. You've just seen a demonstration of two cylinders rolled out of the same amount of material.

    a. Describe what you found.

    _____

    _____

    _____

    _____

    b. How can you explain this result?

    _____

    _____

    _____

    _____

Teacher Page

2. RESEARCH THE PROBLEM: WE NEED WATER!
RESEARCH PHASE 3: CYLINDERS

OBJECTIVES
Students will:

- use a ruler to measure three-dimensional objects (cylinders)
- calculate the surface area and volume of three-dimensional objects
- analyze a table of values for patterns
- make and test conjectures about the relationship between surface area and volume, and dimensions and volume

MATERIALS
For each team:
- ruler
- calculator
- container of dry beans
- graduated cylinder with mL markings
- funnel
- one cylinder from Appendix B

For each class:
- poster board for the cylinders
- Velcro strips for cylinders

BEFORE YOU TEACH
- Prepare the data table shown on page 64 on chart paper or on the board.
- Make a set of cylinders using the patterns in Appendix B. Duplicate some cylinders if there are more than six teams.

OPTIONAL CCSS ENHANCEMENT
To address additional aspects of the Common Core State Standards, demonstrate the derivation of the relationship between the circumference and area of a circle with the given data.

Teacher Page (continued)

1. **CLASS**

   ASK THE CLASS:

   - Were the surface areas of the two cylinders from Research Phase 2 really the same?
     Possible answer(s): No, because the tabletop was acting as the base of the cylinder, thus forming another surface, and the bases of the two cylinders are not the same. If we were to make the bottoms of the cylinders, the shorter/wider cylinder would actually require more material to make than the taller/narrower cylinder.

   - So if the amount of material to make two different cylinders with bottoms was really the same, do you think that these two cylinders would have the same or different volumes?
     Possible answer(s): At this point, some students might still think that the same amount of material (surface area) equals the same volume, while others might think that even with the same surface area, other factors such as the shape of the cylinder would change the volume.

2. **TEAMS** Give each team one cylindrical can. Students will measure the height and radius, calculate the surface area of their cylinder (see table in the answer key), and share their results with the class. To calculate the surface area, students can unwrap the cylinder to show that it is made up of a circle base and a rectangle surface. Students can calculate the area of the circle using the formula πr^2, find the area of the rectangle using length • width, and add the two results together to get the total surface area. Demonstrate how the width of the rectangle is also the circumference of the circle. Teams that finish early can trade their cylinders and check each other's measurements and calculations. All teams should have the same surface area ($\approx$236 cm^2). If they don't, discuss how human error in measurement may have affected the results.

   Optional: Ask students to find the ratio of $\dfrac{\text{circumference}}{\text{diameter}}$. The number is approximately π.

3. **TEAMS** Give teams 2 minutes to answer question 2, where they discuss and predict which can will hold the least amount of beans and which can will hold the most amount of beans and why.

4. **CLASS** Ask students to share their predictions and reasoning with the rest of the class.

5. **TEAMS** Instruct teams to measure the volume of their cylinder using the beans and a graduated cylinder. They should record the volume in question 3 and be ready to share their results with the class.

2. RESEARCH THE PROBLEM: WE NEED WATER!

RESEARCH PHASE 3: CYLINDERS

1. Your group has been given one cylinder. Letter: _____

   a. Measure the radius of your cylinder: _____ cm

   b. Measure the height of your cylinder: _____ cm

   c. "Unwrap" your cylinder, and calculate its total surface area.

Show all calculations.

Can	Radius (cm)	Height (cm)	Circumference (cm)	Surface Area (sq. cm or cm²)	Volume (mL)

STOP   **Wait for your teacher to tell you to continue.**

ANSWER QUESTION 2 AFTER EACH GROUP HAS REPORTED THE MEASUREMENTS OF THEIR CYLINDER.

2. Do you think the cans will all hold the same amount of water? _____

   If you said "yes," explain why.

   _____

   _____

   If you said "no," explain which of the cans you think will hold the most, and why.

   _____

   _____

3. Measure the volume of your cylinder by filling your can to the top with dry beans. Report your data to the class.

   Volume = _____ mL

   Optional: $\dfrac{\text{circumference}}{\text{diameter}}$ = _____

2. RESEARCH THE PROBLEM: WE NEED WATER!
RESEARCH PHASE 3: CYLINDERS

OBJECTIVE: Students will make a double-line graph.

MATERIALS

For each team:
- ruler
- colored pencils or markers (2 different colors)

BEFORE YOU TEACH
- Copy the graph on a transparency.

CLASS OR TEAMS OR INDIVIDUALS Create a double-line graph to show height versus volume and radius versus volume using the class data table.

OPTIONAL CCSS ENHANCEMENTS

To address additional aspects of the Common Core State Standards, direct students to use the terms "independent" and "dependent variable" and to identify the independent and dependent variables in the class data. In this exercise, volume depends on how we change the cylinder dimensions.

2. RESEARCH THE PROBLEM: WE NEED WATER!
RESEARCH PHASE 3: CYLINDERS

4. Use your cylinder measurements to create a double-line graph to show radius versus volume and height versus volume. Use a different color for each line. Color in the key for height and radius.

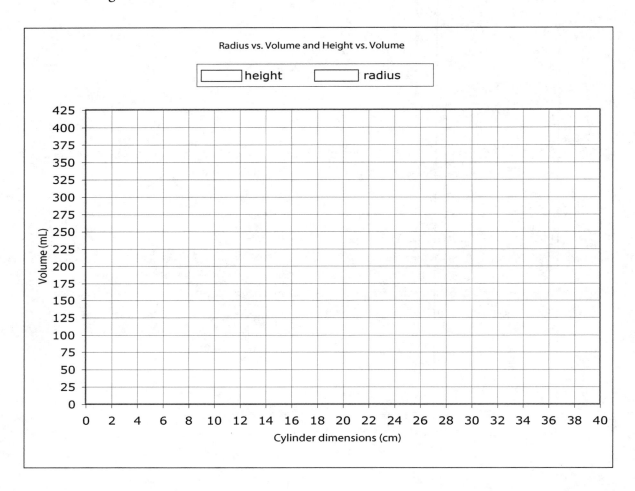

© Museum of Science (Boston), Wong, Brizuela

Teacher Page

2. RESEARCH THE PROBLEM: WE NEED WATER!
RESEARCH PHASE 3: CYLINDERS

OBJECTIVE

Students will:
- analyze the double-line graph to describe the relationship between each pair of variables
- identify the significance of the point where the two lines intersect at the highest point
- connect the results of this research to the engineering design challenge

MATERIALS

For each class:
- The Rules of Thumb list

1.   **TEAMS**   Instruct teams to answer questions 5–8.
   - Questions 5 and 6 ask them about the relationship between radius and height when the surface area is the same. The formula for the surface area of an open cylinder is $\pi r^2 + 2\pi r^2 h$. Since both h and r contribute directly to the surface area, they vary inversely when surface area is held constant.
   - Questions 7 and 8 help students generalize that the volume of the cylinder is greatest, given the same surface area, when the radius equals the height or the appearance that the width (diameter) is double the height.

2.   **CLASS**   Discuss the answers to the questions. The first set of cans provides students with one example of the fact that when the radius equals height, the volume is maximized (for open cylinders with the same surface area). However, one example is not a proof. Students may still be skeptical about this relationship.

   ASK THE CLASS:
   - Do you think the volume will always be the greatest when radius equals height in a cylinder of a certain surface area? Or was this cylinder just a special case?
     Possible answer(s): If students think that this is just a special case and they aren't convinced about this "rule" to maximize the volume of a cylinder, complete the optional activity on pages 69–70.

3.   **CLASS**   Review the design challenge, criteria, and constraints, and ask students what they learned from this research phase that they can add to the Rules of Thumb list.

RESEARCH

2. RESEARCH THE PROBLEM: WE NEED WATER!

RESEARCH PHASE 3: CYLINDERS

Answer questions 5–8 after each group has reported their volume data and plotted their data on the graph.

5.   As the radius increases, what happens to the height?

   _____

   _____

   _____

6.   As the height increases, what happens to the radius?

   _____

   _____

   _____

7.   Which can has the largest volume?_____
   Do you notice anything interesting about this can's radius and height?

   _____

   _____

   _____

8.   How would you create a cylinder with the largest possible volume, if you have a limited amount of material to make the cylinder?

   _____

   _____

   _____

Teacher Page

2. RESEARCH THE PROBLEM: WE NEED WATER!

MORE CYLINDERS (*OPTIONAL ACTIVITY*)

Do this activity if students need more convincing about how to maximize volume given a fixed surface area. Students repeat the cylinders activity with different sets of cylinders.

OBJECTIVES

Students will:

- use a ruler to measure three-dimensional objects (cylinders)
- calculate the surface area and volume of three-dimensional objects
- analyze a table of values for patterns
- make and test conjectures about the relationship between surface area and volume, and dimensions and volume

MATERIALS

For each team:

- calculator
- container of dry beans
- graduated cylinder with mL markings
- funnel
- one cylinder net from one color set from Appendix C (do not mix sets)

For each class:

- Rules of Thumb list
- poster board to make cylinders (orange, pink, and yellow)

BEFORE YOU TEACH

- Prepare blank data table(s) from the table shown below on chart paper or on the board.
- Prepare cylinder nets according to the directions in Appendix C. You don't need to make all three sets if you are limited in time or if you feel that students don't need much more convincing about the rule for maximizing the volume of a cylinder. Also, make duplicate cylinders within each set so that you have enough cylinders (one per team).
- Data table:

Color: _____ Surface area: _____					
	A	B	C	D	E
Radius (cm)					
Height (cm)					
Volume (cm^3)					

1. **TEAMS** Give each team one cylinder from the same color set. Do one set at a time and do not mix sets. You may either choose to tell students that each cylinder within a set has the same surface area, or ask them to calculate the surface area using the radius and height measurements. To save time, give students the radius and height measurements from the table below. For each set, instruct students to fill the cans with beans and then pour the contents of each can into a graduated cylinder. As a class, record and compare the volumes of the cans and determine which can holds the most. For each set, the can for which the radius equals height will have the greatest volume.

   Radius and height for each set of cylinders:

Orange set: surface area ≈ 85 cm²					
	A	B	C	D	E
Radius (cm)	1.5	2	3	4	4.5
Height (cm)	8.3	5.8	3	1.8	0.8

Pink set: surface area ≈ 115 cm²					
	A	B	C	D	E
Radius (cm)	1.5	2.5	3.5	4.5	5
Height (cm)	11.5	6.1	3.5	1.8	1.2

Yellow set: surface area ≈ 151 cm²					
	A	B	C	D	E
Radius (cm)	2	3	4	5	5.5
Height (cm)	11	6.5	4	2.3	1.6

2. **CLASS** Conclude with a class discussion about patterns students notice in the data. Ask: "Are you convinced that volume is greatest when radius equals height for a set surface area?" Then have students discuss how they would create a cylinder that could hold the most water, if they had a limited amount of material.

   For a more specific answer, if surface area = x (represents a limited supply of material),

   $$x = \pi r^2 + 2\pi rh \qquad \text{if } r = h \text{ (radius = height), let } y = r = h$$
   $$x = \pi y^2 + 2\pi y^2$$
   $$x = 3\pi y^2$$

   The radius and height should equal the square root of the quotient of surface area divided by 3π.

   $$y = \sqrt{\frac{x}{3\pi}}$$

3. **CLASS** Review the design challenge, criteria, and constraints, and ask students what they learned from this research phase that they can add to the Rules of Thumb list.

2. RESEARCH THE PROBLEM: WE NEED WATER!
RESEARCH PHASE 4: SQUARE BOXES

OBJECTIVES

Students will:
- use a ruler to measure three-dimensional objects (prisms with square bases)
- calculate the surface area and volume of three-dimensional objects
- analyze a table of values for patterns
- make and test conjectures about the relationship between surface area and volume, and dimensions and volume

MATERIALS

For each team:
- ruler
- calculator
- 1 box

For each class:
- poster board for making boxes

BEFORE YOU TEACH
- Make boxes by following the directions in Appendix D. Make duplicates as needed so each team gets a box.
- Prepare a blank data table from the one below on chart paper or on the board.

CLASS DATA TABLE

Base side (length & width) (cm)	Height (cm)	Surface area (sq. cm or cm^2)	Volume (mL) = width • depth • height
3	19.5	≈243	175.5
5	10.9	≈243	272.5
7	6.9	≈243	338.1
9	4.5	≈243	364.5
11	2.8	≈243	338.8
13	1.4	≈243	236.6

1. **CLASS** Ask students to consider the boxes that they will be investigating in this activity and ask this question: "If several open boxes are all made out of the same amount of material, will they have the same volume?" Some students might say "no" because of the cylinder activity. But others might seem uncertain because of the different shape.

2. **TEAMS** Give each team a box and instruct them to measure the dimensions of the box and calculate its surface area. They should record their results in problem 1.

3. **CLASS** Post each team's results on the data table. Ask the class: "Which of these boxes do you think will hold the least? Which do you think will hold the most? How can you apply what you learned about cylinders in the last activity to help you decide?" Students might say that since tall, narrow cylinders didn't hold as much as wide, short cylinders, the same might be true of boxes. Thus, the short, wide boxes should hold more than the taller, narrow boxes.

4. **TEAMS** Instruct each team to calculate the volume for their box and record it in problem 2.

5. **CLASS** Post each team's results on the data table.

RESEARCH

2. RESEARCH THE PROBLEM: WE NEED WATER!

RESEARCH PHASE 4: SQUARE BOXES

1. Your group has been given one box.

    a. Measure the width of your box: _____ cm

    b. Measure the depth of your box: _____ cm

    c. Measure the height of your box: _____ cm

    d. "Unwrap" your box and calculate its total surface area.

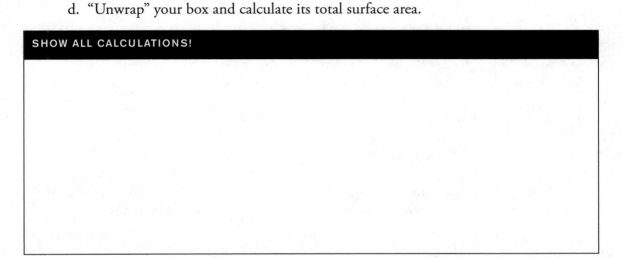

SHOW ALL CALCULATIONS!

2. Calculate the volume of your box using the following formula: Volume = $w \cdot d \cdot h$

SHOW ALL CALCULATIONS!

Volume = _____ cubic centimeters (cm^3)

(*Note:* 1 cubic centimeter (cm^3) holds 1 mL of water.)

STOP   **Wait for your teacher to tell you to continue.**

Teacher Page

2. RESEARCH THE PROBLEM: WE NEED WATER!
RESEARCH PHASE 4: SQUARE BOXES

OBJECTIVES
Students will:
- draw conclusions based on their analysis of the patterns in the class data table
- compare the results of this activity to the results of the previous activity on cylinders
- connect the results of this research to the engineering design challenge
- apply the results of this research to a new 3-D shape

MATERIALS
For each class:
- the Rules of Thumb list

1. **TEAMS** Instruct teams to answer questions 3–6.

2. **CLASS** Go over questions 3–6.

3. **CLASS** Review the design challenge, criteria, and constraints, and ask students what they learned from this research phase that they can add to the Rules of Thumb list.

EXTENSION
Note that the largest cylinder, with a surface area of 236 cm^2, can hold 393 mL. The largest box, with a surface area of 243 cm^2, can hold 364.5 mL. If you have a limited amount of material, would you rather make a cylinder or a box? Why?

ANSWER
The cylinder is a better choice because it can hold more and has a lower surface area (it requires less material). But one should take into account other considerations such as what the packages are designed to contain, the ease of packing and transporting, and the preference of the consumer.

---

2. RESEARCH THE PROBLEM: WE NEED WATER!
RESEARCH PHASE 4: SQUARE BOXES

Answer questions 3–6 after each group has reported their volume data.

3. Which box had the largest volume? _____
    Look at the dimensions of this box. Do you notice anything interesting?

    _____

    _____

    _____

4. Based on the class data, how do you think you would create a box with the largest possible volume if you had a limited amount of material to make it?

    _____

    _____

    _____

5. Compare your answer to question 4 with your answer to question 8 from Research Phase 3: Cylinders. Do you notice anything similar about how to maximize the volume of a box versus the volume of a cylinder? Explain your answer.

    _____

    _____

    _____

6. Based on your research, how do you think you would maximize the volume of a cup that has a hexagonal base (see figure below) if you have a limited amount of surface area?

    _____

    _____

    _____

3. BRAINSTORM POSSIBLE SOLUTIONS: WE NEED WATER!
RESEARCH PHASE 3: CYLINDERS

OBJECTIVE: Students will individually brainstorm and sketch a water collector design.

MATERIALS

For each class:
- the Rules of Thumb list

1. **CLASS** Instruct students to answer question 1 together. The area of the plane siding is 1987 square centimeters (cm²). Review the design criteria with students, and discuss other factors they should consider when designing the water collector: Should it have a handle for easy pouring? A cover to keep out bugs and dirt when it's not raining? An easy-pour spout?

2. **CLASS** Review the Rules of Thumb list for any useful suggestions to meet the design criteria and constraints.

3. **INDIVIDUALS** Instruct students to individually brainstorm ideas for using the given area of the material to figure out the shape and dimensions of the water collector. If appropriate, they may finish their brainstorm and sketch for homework.

INTERESTING INFO

The Nabataeans were a trading people of ancient Arabia. They were experts at collecting water and storing it in underground cisterns. The Nabataeans' greatest accomplishment was probably their system of water management. They developed a system to collect rainwater using water channels, pipes, and underground cisterns. They also developed sophisticated ceramic pipelines using gravity feeds that served the developing urban centers. They hid these systems all along their caravan routes to have water available to them but not to other people.

BRAINSTORM

3. BRAINSTORM POSSIBLE SOLUTIONS: WE NEED WATER!

INDIVIDUAL DESIGN

ENGINEERING CRITERIA

STRONG → It remains intact when filled to capacity.

LARGE CAPACITY → When filled, it does not leak. This means that it should have a bottom.

LEAKPROOF → It can hold at least 1,000 mL of water for each member of your team, and the more the better!

FREESTANDING → It can stand upright without anyone holding it.

ENGINEERING CONSTRAINTS

1. In Research Phase 1, you determined how many square centimeters of plane siding you have. Write this value here: _____ square centimeters (cm²). Remember, this is all the material you have.

2. Brainstorm independently. Sketch out one possible water collector design in the space below. Figure out what dimensions (height, diameter, and so forth) would work given the available amount of material. Label all parts. Be creative!

Teacher Page

4. CHOOSE THE BEST SOLUTION: WE NEED WATER!

OBJECTIVE: Students will share their individual brainstorm designs and decide on a team design.

MATERIALS

For each team:
- a poster board of the plane siding

1. **TEAMS** Teams should share their brainstorm designs and decide on their final team design. Give each team the poster board of the plane siding to manipulate as they decide on their design. They may not fold or cut the plane siding until the next EDP step. To encourage students to apply their research results, you may want to tell teams that you will award a prize to the team that designs and builds a water collector with the largest volume.

CLASSROOM MANAGEMENT TIPS
- If students need more structure to share their individual water collector designs from the Brainstorm step, go over a list of things for each person to share (for example, the shape of the collector, the rationale for shape, the dimensions of the collector, the calculated volume, the amount of materials used, how the collector would be constructed, and so forth). Establish a set time for each person to share (1 minute) and behavior expectations for teammates during sharing (for example, no interruptions, no comments until everyone has shared, active listening).
- Discuss with students how to come to an agreement on a team design. Team members can take turns discussing pros and cons of each design; and can identify commonalities in designs, compromise on areas of disagreement, and either vote or try to reach consensus.

---

2. **CLASS** If you plan on using students' team designs as an assessment, review the rubric on page 129 with students.

ASSESSMENT

Use the Rubric for Engineering Drawings on page 129 to assess students' work.

INTERESTING INFO

Being able to sketch is an important ability to have in engineering. You will often need to make diagrams and keep track of measurements in projects. Here are some tips for sketching.
- Draw using your shoulder rather than your wrist. Drawing with the shoulder allows you to make continuous lines in one straight movement. With less wrist movement, there will be fewer jagged edges and bumps.
- Use a pad of layout paper. Layout paper has very light gridlines.
- Use a pencil.
- Draw a proportion scale in your sketch.

Teacher Page

4. CHOOSE THE BEST SOLUTION: WE NEED WATER!

OBJECTIVE: Students will calculate the surface area (and perhaps volume) of their team designs.

1. **TEAMS** Teams should also complete question 2, which shows their calculations to make sure that they have enough materials to build the prototype according to the dimensions they used in their design. If possible, challenge students to also calculate the volume of their design, and ask them how they can adjust the dimensions to optimize the volume.

2. **TEAMS** After teams finish their design, check to make sure that their drawing is clearly labeled with dimensions. If their drawing and calculations are complete, then they may begin building their prototype. If not, they need to continue working on the design and calculations until they are complete.

INTERESTING INFO: Math Problem

Suppose that you want to construct a plastic cube with a volume of 125 cm^3 to hold water. You need a special sealant along all the edges of the cube. You have enough sealant to cover 40 cm of the cube's edges. Identify the engineering criteria and constraint. Is it possible to meet the criteria and constraint?

ANSWER: No. The engineering criteria is to create a sealed plastic cube with a volume of 125 cm^3 to hold water. The engineering constraint is that you only have 40 cm of sealant. You cannot seal the cube because of the engineering constraint. To find the length of an edge of a cube with a volume of 125 cm^3, find the cubic root of 125, which is 5, because $5 \cdot 5 \cdot 5 = 125$. A cube has 12 edges, so you will need $12 \cdot 5$ or 60 cm of sealant. You only have 40 cm of sealant, so you cannot meet the engineering criteria of sealing the entire cube.

4. CHOOSE THE BEST SOLUTION: WE NEED WATER!

TEAM DESIGN

1. Discuss your ideas as a group. Decide on a team design that meets the engineering criteria and constraints. Then sketch your design in the space below. Please label the dimensions (height, diameter, radius, and so forth) on your diagram. Use the Rubric for Engineering Drawings provided by your teacher to check the quality and completeness of your drawing.

2. Do you have enough material to make your water collector?   ☐ yes   ☐ no

Use the dimensions that you labeled on your design sketch in question 1 to calculate the total square centimeters (cm²) of material needed to construct your water collector. Show your work!

HELPFUL SURFACE AREA FORMULAS ($\pi \approx 3.14$)

Open cylinder $= \pi \cdot r^2 + 2 \cdot \pi \cdot r \cdot h$

h

r

Open rectangular box $= w \cdot d + 2(w \cdot h) + 2(d \cdot h)$

h

w

d

Open cone $= \pi \cdot r \cdot s$

r

s

WORKSPACE

Teacher Page

5. BUILD A PROTOTYPE/MODEL: WE NEED WATER!

OBJECTIVE: Students will build their teams' water collector design.

MATERIALS

For each team:
- poster board of plane siding
- tape
- scissors

TEAMS As teams begin building their prototype, instruct them to write down or draw any changes to the original design.

ASSESSMENT

Use the Rubric for Prototype/Model on page 131 to assess completeness and craftsmanship of students' models.

TIP

Warn teams that they only have the materials they're given. If students have problems for whatever reason (for example, they run out of materials, something gets torn, or materials are measured incorrectly), they cannot get replacement materials, just as if they were stranded in real life. They will just have to make do with whatever materials they have left. Make sure that they record any changes to their design.

5. BUILD A PROTOTYPE/MODEL: WE NEED WATER!

Collect all necessary materials and build your model according to the solution you chose in step 4. As you build your model, write down any changes to your original design. Use the Rubric for Prototype/Model provided by your teacher to assess your work.

6. TEST YOUR SOLUTION: WE NEED WATER!

OBJECTIVE: Students will follow testing procedures to test their water collector prototypes.

MATERIALS
For each team:
- dry beans
- graduated cylinder
- water collector model

1. | **CLASS** | Have each team follow the directions to test their design.

2. | **TEAMS** | After completing the tests, instruct teams to answer questions 1 and 2.

ASSESSMENT
Use the Rubric for Test, Communicate, and Redesign Steps on page 132 to assess how well students followed testing procedures.

TEST

6. TEST YOUR SOLUTION: WE NEED WATER!

Perform the following tests on your model. For each test, check "yes" if your water collector passes the test and "no" if it fails. You may use the Rubric for Test, Communicate, and Redesign Steps provided by your teacher to assess your work on the next few pages.

STRENGTH TEST

Fill your water collector to the top with dry beans. Does your collector stay intact?

☐ yes   ☐ no

LEAKPROOF TEST

Fill your water collector to the top with dry beans. Do all the beans remain inside the filter (no leaks)?

☐ yes   ☐ no

FREESTANDING TEST

Does your water collector stand upright by itself?

☐ yes   ☐ no

CAPACITY TEST

Use the graduated cylinder to measure the beans held by your collector. Does your collector hold at least 1,000 mL of beans (water) per team member?

☐ yes   ☐ no

What is the exact volume of your collector? _____ cm^3

EVALUATE YOUR DESIGN

1. Based on the four tests, how well did your water collector meet the design criteria?

    _____

    _____

    _____

2. What are some areas where you can improve your water collector design?

    _____

    _____

    _____

7. COMMUNICATE YOUR SOLUTION: WE NEED WATER!

OBJECTIVE: Students will answer questions to reflect on their designs, present their designs to another team, and give feedback on the other team's design.

TEAMS Pair teams and instruct them to choose one person to describe their design to the other team, and to share their evaluation from the testing page. Each team member is to offer at least one suggestion to the other team to improve that team's design. Set the time limit to about 15 minutes.

7. COMMUNICATE YOUR SOLUTION: WE NEED WATER!

Teams will now pair up to share their designs, and provide feedback and suggestions.

Nominate one person in your group to describe your model and share your evaluation (from step 6) with the other team.

Each member of the other team should provide one suggestion for how to improve your design.

Please record these suggestions in the space below.

Suggestion 1:

_____

_____

_____

Suggestion 2:

_____

_____

_____

Suggestion 3:

_____

_____

_____

Suggestion 4:

_____

_____

_____

8. REDESIGN AS NEEDED: WE NEED WATER!

OBJECTIVE: Students will answer questions to consider how they can redesign their water collectors.

1.   **TEAMS** Instruct teams to use the suggestions they received from the other team to describe changes they would make to their original design. Set the time limit to about 7 to 8 minutes.

2.   **CLASS** If there is enough time, allow each team a few minutes to show their water collector and give a brief persuasive argument on why it's a great design. They can talk about how well it met the criteria and constraints. Let the class vote on which team's water collector they would most like to have with them on the island.

REDESIGN

8. REDESIGN AS NEEDED: WE NEED WATER!

As a group, review the other team's suggestions.

1. Based on their suggestions, how will you change or improve your design?

    _____

    _____

    _____

    _____

    _____

    _____

    _____

2. The class will vote on which team's water collector they would most like to have with them on the island. Before the vote, describe the necessary improvements, and prepare a persuasive argument about why your team's design should be chosen.

    _____

    _____

    _____

    _____

    _____

    _____

    _____

3. Consider the water collector that was voted on by the class. Why do you think this water collector was chosen over the others?

    _____

    _____

    _____

    _____

    _____

    _____

INDIVIDUAL SELF-ASSESSMENT RUBRIC: WE NEED WATER!

OBJECTIVE: Students will use a rubric to individually assess their involvement and work in this design challenge.

ASSESSMENT

Assign this reflection exercise as homework. You can write your comments on the lines below the self-assessment, and/or use this in conjunction with the Student Participation Rubric on page 133.

Name _____

INDIVIDUAL SELF-ASSESSMENT RUBRIC: WE NEED WATER!

Use this rubric to reflect on how well you met behavior and work expectations during this activity. Check the box next to each expectation that you successfully met.

LEVEL 1	LEVEL 2	LEVEL 3	LEVEL 4	BONUS POINTS
Beginning to meet expectations	Meets some expectations	Meets expectations	Exceeds expectations	
☐ I was willing to work in a group setting.	☐ I met all of the Level 1 requirements.	☐ I met all of the Level 2 requirements.	☐ I met all of the Level 3 requirements.	☐ I helped resolve conflicts on my team.
☐ I was respectful and friendly to my teammates.	☐ I recorded the most essential comments from other group members.	☐ I made sure that my team was on track and doing the tasks for each activity.	☐ I helped my teammates understand the things that they did not understand.	☐ I responded well to criticism.
☐ I listened to my teammates and let them fully voice their opinions.	☐ I read all instructions.	☐ I listened to what my teammates had to say and asked for their opinions throughout the activity.	☐ I was always focused and on task: I didn't need to be reminded to do things; I knew what to do next.	☐ I encouraged everyone on my team to participate.
☐ I made sure we had the materials we needed and knew the tasks that needed to be done.	☐ I wrote down everything that was required for the activity.	☐ I actively gave feedback (by speaking and/or writing) to my team and other teams.	☐ I was able to explain to the class what we learned and did in the activity.	☐ I encouraged my team to persevere when my teammates faced difficulties and wanted to give up.
	☐ I listened to instructions in class and was able to stay on track.	☐ I completed all the assigned homework.		☐ I took advice and recommendations from the teacher about improving team performance and used feedback in team activities.
	☐ I asked questions when I didn't understand something.	☐ I was able to work on my own when the teacher couldn't help me right away.		☐ I worked with my team outside of the classroom to ensure that we could work well in the classroom.
		☐ I completed all the specified tasks for the activity.		

Approximate your level based on the number of checked boxes: _____ Bonus points: _____

Teacher comments: _____

TEAM EVALUATION: WE NEED WATER!

OBJECTIVE: Students will evaluate and discuss how well they worked in teams.

ASSESSMENT

1. **INDIVIDUALS** Assign this reflection exercise as homework or during quiet classroom time.

2. **TEAMS** Instruct students to share their team evaluation reflections with one another and discuss how they can improve their teamwork during the next activity.

3. **CLASS** Point out any good examples of teamwork and areas to improve during the next activity.

TEAM EVALUATION: WE NEED WATER!

How well did your team work together to complete the design challenge? Reflect on your teamwork experience by completing this evaluation and sharing your thoughts with your team. Celebrate your successes and discuss how you can improve your teamwork during the next design challenge.

RATE YOUR TEAMWORK. On a scale of 0–3, how well did your team do? 3 is **excellent;** 0 is **very poor.** Explain how you came up with that rating. Was it the same, better, or worse than the shelter activity?

LIST THINGS THAT WORKED WELL. Example: We got to our tasks right away and stayed on track.

LIST THINGS THAT DID NOT WORK WELL. Example: We argued a lot and did not come to a decision that everyone could agree on.

HOW CAN YOU IMPROVE TEAMWORK? Make the action steps concrete. For example: We need to learn how to make decisions better. Therefore, I will listen and respond without raising my voice.

Teacher Page

Design Challenge 3
Balancing Act!

INTRODUCTION

OBJECTIVE: Students will read and understand the problem presented for the third design challenge.

CLASS   Together, read the introduction. Share with the class that the colorful pattern on the cover of the book was inspired by traditional Maori designs.

INTERESTING INFO
The Maori people are indigenous to New Zealand. The word *Maori* means "normal." The Maori often refer to themselves as *tengata whenua*, which means "people of the land." The Maori people traveled the ocean in giant canoes. Using the stars, sun, and sea currents to navigate and find land, they traveled from island to island. They migrated to New Zealand from the Cook Islands, Society Islands, and Marquesas Islands in the South Pacific around 800 c.e.

In 1841, New Zealand became an official British colony. Many European settlements were soon established. Conflicts soon arose between the Maori and European colonizers over claims to the land. These conflicts escalated into the New Zealand Wars in 1860.

After the New Zealand Wars, many Maori lands were confiscated. Remaining Maori lands were generally very poorly suited to farming. Most of the Maori people lived in small rural communities separated from the European settlements. The Maori population declined rapidly as a result of the wars and European diseases to which they had little immunity, such as influenza, measles, and whooping cough. In the late nineteenth century, European settlers referred to the Maoris as a "dying race."

© Museum of Science (Boston), Wong, Brizuela

Design Challenge 3

Balancing Act
INTRODUCTION

You and your stranded classmates have been on this deserted island for days with no hope of being rescued—that is, until now! You spot a group of Maori (pronounced "maw-ree") people canoeing not far in the distance! You wave and scream to get their attention. They spot you and head to the island's shore.

1. DEFINE THE PROBLEM: BALANCING ACT!

OBJECTIVE: Students will read and understand the criteria and constraints of the design challenge.

CLASS Together, read the activity.

ASK THE CLASS:
- "Have you ever been canoeing? If so, have you ever tipped over?
  What causes canoes to tip over? How can we keep a canoe from tipping over?"
  Possible answer(s): Canoes tip over when they are imbalanced—when they have too much weight on one side. Canoes tip over when someone stands up and causes the canoe to rock back and forth too much. To keep a canoe from tipping over, one must be careful when moving from one part of the canoe to another. Also, one must be careful to keep the canoe balanced when putting people or supplies in it.

INTERESTING INFO
The dugout canoe was the principal means of transportation for people living along or near rivers and streams. These canoes varied in size and quality. The várzea chiefdoms—the Tapajós and the Omagua—built large, sturdy canoes for long-distance war expeditions up and down the Amazon River. On one occasion, the Spaniards counted as many as 8,000 warriors in 130 canoes (equal to about 60 men per canoe), which were paddled out to attack the intruding Europeans.

1. DEFINE THE PROBLEM: BALANCING ACT!

The Maori people have agreed to take you and your classmates to New Zealand in their canoe. However, they have some concerns. The trip to New Zealand will take several days, so you will need to bring along food and water. Also, the canoe is extremely unstable. If the canoe is even the slightest bit unbalanced, it will tip over! Therefore, you will have to load the canoe very carefully.

ENGINEERING CRITERIA

ALL OF THE FOLLOWING ITEMS MUST BE LOADED INTO THE CANOE WITHOUT IT TIPPING OVER:

- 9 students
- 3 Maori sailors
- 3 canoeing paddles
- 1 life preserver
- 1 inflatable raft with rope
- food supply (coconuts, bananas, etc.)
- water supply (filled water collectors)

ENGINEERING CONSTRAINTS

- The canoe is 10 meters long.
- The canoe has 40 seats, and consecutive seats are separated by 0.25 meters of space.
- There isn't a seat in the center of the canoe.
- At most, two people or items can be placed at any one seat, totaling 60 kilograms per seat.

A seesaw balance is used to model the canoe (see diagram below). The balance has 10 equally spaced pegs on each lever arm to represent the 20 seats on the canoe.

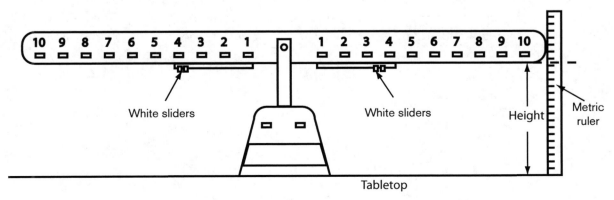

White sliders   White sliders   Height   Metric ruler

Tabletop

Teacher Page

2. RESEARCH THE PROBLEM: BALANCING ACT!
RESEARCH PHASE 1: BALANCING TWO OBJECTS

OBJECTIVES
Students will:
- investigate how the weight and distance of objects on a horizontal platform with a center fulcrum relate physically and mathematically to keep the platform balanced
- generalize and represent a pattern using words or symbols

MATERIALS
For the class:
- transparencies of the student worksheet on page 100 (optional)
- overhead projector (optional)
- transparency markers (optional)
- blank transparencies (optional)

For each team:
- Math Balance kit (available at http://classroomproductswarehouse.com)

1. **CLASS**

   ASK THE CLASS:
   - What are some of your ideas for loading the canoe with these people and supplies so that the canoe won't tip over?

     Possible answer(s): Put half of the people on one side and half on the other. But there are an odd number of students and an odd number of Maori, so it's not quite that easy. Also, different items and people have different weights.

2. **CLASS** Distribute a Math Balance kit to each team and show them how to calibrate the balance. First, make sure that there are no tiles on any of the pegs. Adjust the white sliders until the distance between the bottom edge of the lever's outmost corners and the tabletop is the same for both levers. When perfectly balanced, the height should be approximately 19.8 centimeters.

3. **TEAMS** Instruct teams to follow the directions on page 100 and answer questions 1–4. This should take about 10 minutes. Students will use specific examples to look for patterns and attempt to generalize the relationship between the left and right sides of the seesaw.

4. **CLASS** After going over the solutions for problems 1–3 (on transparency if needed), discuss question 4 in greater depth. After students realize that the weight times distance on the left equals weight times distance on the right, ask the class: Relating this to playing on a seesaw, how would you balance an 18-kilogram child with a 54-kilogram adult?

   Possible answer(s): Put the child farther away from the center, about 1.5 meters away, and put the 54-kilogram adult closer to the center on the other side, about .5 meters away. Thus, $18 \cdot 1.5 = 27 = 54 \cdot .5$, and the seesaw should be balanced. Other answers are possible (for example, put the child 3 meters from the center and the adult 1 m from the center).

5. **CLASS**

   ASK THE CLASS:

   - How could you represent this pattern or relationship with symbols or variables?

     Possible answer(s): $W_L \cdot D_L = W_R \cdot D_R$ where W represents weight, D represents distance from the center, R represents the right side, and L represents the left side.

   OPTIONAL CCSS ENHANCEMENTS
   To address additional aspects of the Common Core State Standards, relate the sides of the balance to a number line with 0 at the fulcrum. Direct students to write expressions using variables to represent the items being balanced on the fulcrum, as well as quantities of distance and weight. Also, relate net force to "combining to make 0." Direct students to represent an unbalanced situation with an inequality.

2. RESEARCH THE PROBLEM: BALANCING ACT!

RESEARCH PHASE 1: BALANCING TWO OBJECTS

You will investigate how to balance different items on a canoe using a seesaw to model the canoe and tiles to model the different items. Calibrate the balance by adjusting the white sliders and measuring the distance between the bottom edge of the lever arms and the tabletop. The height should be approximately 19.8 centimeters (same for both arms) when the balance is level (see the diagram on page 97).

1. Hang 4 tiles on the 4th peg to the left of center (see below). How many tiles would you have to hang on the 8th peg to the right of center in order to perfectly balance the seesaw? Draw your answer in the diagram below.

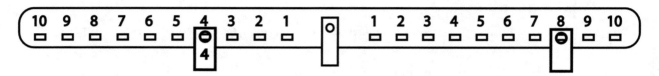

2. Hang 3 tiles on the 8th peg to the left of center (see below). Now, create a stack of 4 tiles. Where would you have to hang the stack of 4 tiles in order to perfectly balance the seesaw? Draw your answer in the diagram below.

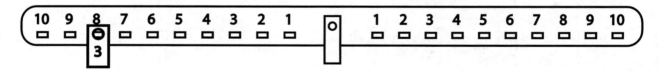

3. Create one stack of 3 tiles, and another stack of 2 tiles. Where can you hang these stacks so that the seesaw will be perfectly balanced? Draw your answer in the diagram below. Can you think of another solution?

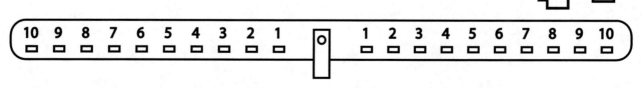

4. Look at your answers to questions 1–3. What patterns do you notice? Do you see a relationship between what is on the left side and the right side of the seesaw?

_____

_____

Teacher Page

2. RESEARCH THE PROBLEM: BALANCING ACT!
RESEARCH PHASE 2: BALANCING MORE THAN TWO OBJECTS

OBJECTIVE: Students will investigate how the weight and distance of objects on a horizontal platform with a center fulcrum relate physically and mathematically to keep the platform balanced.

MATERIALS
For the class:
- transparencies of the student worksheet on page 103 (optional)
- overhead projector (optional)
- transparency markers (optional)
- blank transparencies (optional)

For each team:
- Math Balance kit

1. $\boxed{\textbf{TEAMS}}$ Instruct teams to use their Math Balance kits to model and solve problems 1–4. Remind students to think about what they learned in the first set of problems.

   ASK EACH TEAM: How do these problems relate to the first set of problems? How could what you've already learned help you solve these problems?

EXTENSION 1

It's possible to solve problems 1 and 2 as two-step equations with one unknown per equation. For problem 1, the equation is: $(2 \cdot 8) + (3 \cdot 4) = 7x$, where x represents the number of tiles. For problem 2, the equation is $(1 \cdot 10) + (3 \cdot 2) = 2x$, where x represents the peg position.

---

2. **CLASS**

> **ASK THE CLASS:** What strategies are you using to solve the problems involving more than two objects?
>
> **Possible answer(s):** trial and error, application of relationship discovered with two objects (multiply distance from center and the number of tiles) to more than two objects, and so forth

Problem 3 could take a while if students just use trial and error. Also, the seesaw scale may not be precise enough to show an exact solution; it's possible that the seesaw could appear balanced, but mathematically the sums are not equal. One strategy is to notice that the starting positions of the weights show that the left side ($2 \cdot 4 = 8$) is 5 more than the right side ($3 \cdot 1 = 3$). Thus, in order to make up for the imbalance of 5, the difference of the products of the two stacks that students need to place should also be 5. For instance, in the solution shown in the answer key, the 2 tiles positioned on the right side have a product of 10, which is 5 more than the 1 tile positioned on the left side that has a product of 5.

EXTENSION 2
Encourage advanced students to find all six possible solutions to problem 3. Is there a systematic way to find all the possible solutions?

EXTENSION 3
Encourage students who finish early to make up their own balancing problems that they can use to challenge other students on their team or another team.

Name

STUDENT PAGE

RESEARCH

2. RESEARCH THE PROBLEM: BALANCING ACT!

RESEARCH PHASE 2: BALANCING MORE THAN TWO OBJECTS

1. Hang 2 tiles on the 8th peg to the left of center and 3 tiles on the 4th peg to the left of center. How many tiles would you have to place on the 7th peg to the right of center in order to balance the seesaw perfectly? Draw your answer on the diagram below.

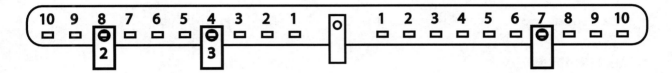

2. Hang 1 tile on the 10th peg to the left of center, 3 tiles on the 2nd peg to the left of center, and 4 tiles on the 1st peg to the right of center. Now, create a stack of 2 tiles. Where would you have to hang the stack of 2 tiles in order to perfectly balance the seesaw? Draw your answer in the diagram below.

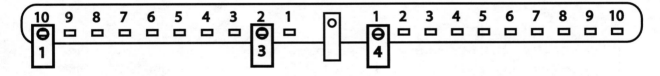

3. Hang 2 tiles on the 4th peg to the left of center and 3 tiles on the 1st peg to the right of center. Now, create a stack of 1 tile and a stack of 2 tiles. Where can you place these stacks so that the seesaw will be perfectly balanced? Draw your answer in the diagram below. Can you think of another solution?

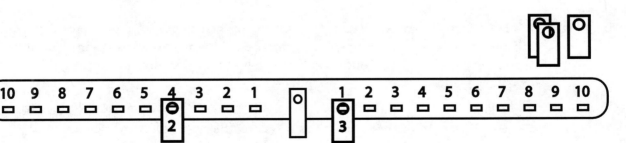

4. Explain your strategy for finding the solution(s) to problem 3.

_____

_____

© Museum of Science (Boston), Wong, Brizuela

Stranded!

103

2. RESEARCH THE PROBLEM: BALANCING ACT!
RESEARCH PHASE 2: BALANCING MORE THAN TWO OBJECTS

OBJECTIVES

Students will:
- investigate how the weight and distance of objects on a horizontal platform with a center fulcrum relate physically and mathematically to keep the platform balanced
- generalize and represent a pattern using words or symbols
- connect the results of this research phase to solving the design challenge

MATERIALS

For the class:
- transparencies of the student worksheet on the next page (optional)
- overhead projector (optional)
- transparency markers (optional)
- blank transparencies (optional)
- a piece of chart paper for the Rules of Thumb list (see page 10)

For each team:
- Math Balance kit

1. **CLASS** Discuss problems 7 and 8 together. You want students to generalize the relationship between the left and right sides of the seesaw. Summation notation may be new to students.

   ASK THE CLASS: Why would we want to use this notation?

   Possible answer(s): Using the notation is a simple and clear way to represent an unknown number of weights on both sides of the equation.

2. **TEAMS** Instruct teams to exchange one copy of their solutions to problems 1–5 with another team's so they can check each other's solutions. After checking solutions, instruct teams to get their work back, examine how well they did, and correct any errors they made.

3. **CLASS** Wrap up the research phases by asking the class: "How can you apply what you learned today to designing a loading plan for the Maori canoe?" Use students' responses to make a Rules of Thumb list.

Name

STUDENT PAGE

RESEARCH

2. RESEARCH THE PROBLEM: BALANCING ACT!

RESEARCH PHASE 2: BALANCING MORE THAN TWO OBJECTS

5.   Create a stack of 1 tile, a stack of 2 tiles, a stack of 3 tiles, and a stack of 4 tiles (see below). Where could you hang these four stacks of tiles in order to perfectly balance the seesaw? Try to come up with two different solutions, and draw your answers on the diagrams below.

Solution 1:

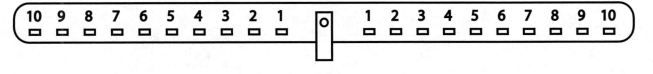

Solution 2:

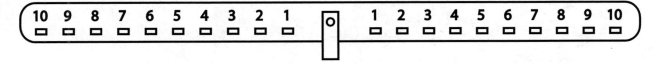

6.   Explain your strategy for finding solutions to problem 5.

_____

_____

7.   Look at your answers to questions 1–4 from the first part of this activity. What patterns do you notice? Do you see a relationship between the left and right sides of the seesaw?

_____

_____

8.   Using pictures or symbols, show how the weights and distances on the left side of the seesaw are related to the weights and distances on the right side of the seesaw.

Teacher Page

3. BRAINSTORM POSSIBLE SOLUTIONS: BALANCING ACT!

OBJECTIVE: Students will review the design challenge's criteria and constraints.

MATERIALS
For the class:
- Rules of Thumb list on chart paper

CLASS   Instruct students to complete Table 3.1. Together, read the Additional Information box and discuss question 1 as a class.

You can elaborate on the question by asking: Should all of the water and food be spread out or in a central location? Does it matter where you put the paddles—do they need to be on a seat next to a person?

Add your answers to question 1 to the Rules of Thumb list.

CANOEING TIPS
- Standing up or moving in a canoe greatly increases the chances of capsizing.
- Always maintain three points of contact with the canoe while moving around.
- Load the boat properly by evenly distributing the weight.
- Keep your shoulders inside the canoe's gunwales (top edges of the sides of the boat).
- Take hands-on training.
- Wear a life jacket.
- Understand the canoe's and your own limitations.
- Plan ahead.
- Know how to swim.
- Never paddle alone.

---

3. BRAINSTORM POSSIBLE SOLUTIONS: BALANCING ACT!

Now that you have figured out how to balance multiple weights on a seesaw, you are ready to design a loading plan for the Maori sailors' very tipsy canoe.

TABLE 3.1 WEIGHTS OF ITEMS TO LOAD

ITEMS TO LOAD ON CANOE	AVERAGE WEIGHT PER ITEM (in kilograms)	WEIGHT IN TILES PER ITEM (Let 1 tile = 15 kg)
9 students	45 kg per student	1. _____ tiles per student
3 Maori people	60 kg per Maori	2. _____ tiles per Maori
3 canoeing paddles	15 kg per paddle	3. _____ tiles per paddle
1 life preserver	15 kg per life preserver	4. _____ tiles per life preserver
1 inflatable raft with rope	30 kg per raft	5. _____ tiles per raft
1 bag of food	30 kg per food bag	6. _____ tiles per food bag
2 filled water collectors	15 kg per water collector	7. _____ tiles per water collector

ADDITIONAL INFORMATION

- The actual canoe is 10 meters long.
- Each position (represented by equally spaced lines) along the canoe can comfortably hold up to 2 people/items totaling 60 kg. For example, you can only place 1 Maori (4 tiles) in a seat, but you can place 1 student (3 tiles) and 1 canoe paddle (1 tile) in the same seat.

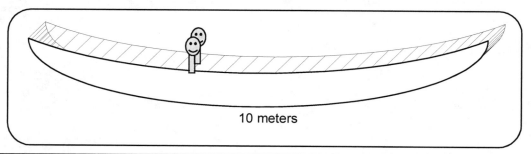

10 meters

8. What else should you take into consideration while you design your loading plan?

_____

_____

_____

Teacher Page

3. BRAINSTORM POSSIBLE SOLUTIONS: BALANCING ACT!

OBJECTIVE: Students will individually brainstorm solutions for balancing pairs of items/people.

MATERIALS

For each team:
- Math Balance kit

1. [**INDIVIDUALS**] Instruct students to individually complete problems 2–4. They should first apply their knowledge of the equation or relationship between the two sides of the seesaw from the research phase. Then they can check their solutions using the seesaw model. Remind students that there are many correct solutions. Challenge advanced students to find as many solutions as possible.

2. [**TEAMS**] Instruct team members to check each other's work. After getting their work back, students should correct any errors.

3. BRAINSTORM POSSIBLE SOLUTIONS: BALANCING ACT!

On your own, you will figure out pairs of items that can be loaded into the canoe at the same time. On the diagrams below, draw where you would place each pair of items in order to keep the canoe perfectly balanced.

9. Balance 1 Maori person with 1 student. [O M 4] [O S 3]

| 10 | 9 | 8 | 7 | 6 | 5 | 4 | 3 | 2 | 1 | O | 1 | 2 | 3 | 4 | 5 | 6 | 7 | 8 | 9 | 10 |

How did you decide where to place the two items?

_____

_____

10. Balance 1 food bag with 1 water collector. [O F 2] [O W 1]

| 10 | 9 | 8 | 7 | 6 | 5 | 4 | 3 | 2 | 1 | O | 1 | 2 | 3 | 4 | 5 | 6 | 7 | 8 | 9 | 10 |

How did you decide where to place the two items?

_____

_____

11. Balance any two items of your choice.

| 10 | 9 | 8 | 7 | 6 | 5 | 4 | 3 | 2 | 1 | O | 1 | 2 | 3 | 4 | 5 | 6 | 7 | 8 | 9 | 10 |

How did you decide where to place the two items?

_____

_____

4. CHOOSE THE BEST SOLUTION: BALANCING ACT!

OBJECTIVE: Students will share their individual brainstorm solutions and work on a team loading plan.

MATERIALS

For the class:
- Rules of Thumb list on chart paper

TEAMS Instruct team members to work together to create a loading plan for the canoe. Encourage them to use their individual brainstorm solutions to problems 2–4 in their plan. They should write out the steps in order and label the diagram to show where the different people and objects should be placed. They should check their solution by showing how the summation equation is true for their final configuration. Remind students to review the Rules of Thumb list when making design decisions.

EXTENSION

Because each seat can hold up to 2 items or individuals, this is a good opportunity to teach the distributive property in a meaningful way. As an example, let's say that you put 1 student (3 tiles) and 1 paddle (1 tile) in the same seat at peg 4. Show that $(3 \cdot 4) + (1 \cdot 4)$ is the same as $(3 + 1) \cdot 4$ or $4 \cdot 4$. After showing students a few more examples, you can generalize the property using variables: $(a \cdot b) + (c \cdot b) = (a + c) \cdot b$.

ASSESSMENT

Introduce students to the Rubric for Canoe-Loading Plan on page 130 by showing students examples of student work on pages 134–141, and using the rubric to grade each drawing. Note that the drawings show a different balance (there are 40 seats, blocks are used instead of tiles, and some items/people have different weights), but they can still be assessed using the same rubric. You can first assess one drawing with the whole class using the rubric, and then have students work in pairs to assess the other drawings. Debrief as a whole class. Students can use the rubric to self-assess their own drawings of the Canoe-Loading Plan design.

4. CHOOSE THE BEST SOLUTION: BALANCING ACT!

Now that you have paired together items/people that can be safely loaded onto the canoe, you are ready to design a step-by-step loading plan. On the model seesaw below, label where each item should be loaded. You must load items in pairs at the same time to keep the canoe balanced. Please number the items in the order that they should be loaded. Show mathematically how the canoe should be balanced. Use the Rubric for Canoe-Loading Plan provided by your teacher to check the quality and completeness of your drawing. Practice using the rubric on the sample drawings provided by your teacher.

Labeling Key

S = student

M = Maori

P = paddle

L = life preserver

R = raft

F = food

W = water

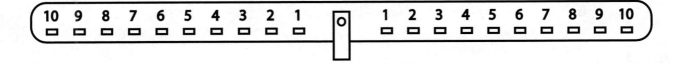

5. BUILD A PROTOTYPE/MODEL: BALANCING ACT!

OBJECTIVE: Students will create and organize stacks of tiles that model people/items to prepare to test their plans.

MATERIALS

For each team:
- Math Balance kit with 49 tiles

Note: Since each Math Balance kit contains only 20 weights, combine teams so that each team has enough weights for the activity.

TEAMS For this activity, instruct teams to use the checklist on the next page to create stacks of tiles that would represent each item or person, and organize them in the order that they will be loaded onto the seesaw.

INTERESTING INFO

On July 1, 1940, the Tacoma Narrows was opened to the public. The Tacoma Narrows was a suspension bridge, which means that only wires held up the bridge. A few months later, on November 7, 1940, the Tacoma Narrows collapsed due to wind-induced resonance.

The Tacoma Narrows collapsed because engineers did not do enough testing on their prototype models. They never considered the possibility that the oscillations in the bridge could cause the collapse. If they had performed careful prototype testing, the disaster could have been averted.

BUILD

5. BUILD A PROTOTYPE/MODEL: BALANCING ACT!

You won't physically "build" anything since you are designing a plan. However, during this step, you can prepare and assemble things for your design test.

1. Create stacks of tiles to represent each of the items to be loaded (see checklist below).

    ☐ 9 students

    ☐ 3 Maori

    ☐ 3 paddles

    ☐ 1 life preserver

    ☐ 1 raft

    ☐ 1 food bag

    ☐ 2 filled water collectors

2. Organize the stacks in the order that they will be loaded onto the seesaw.

6. TEST YOUR SOLUTION: BALANCING ACT!

OBJECTIVE: Students will test their teams' loading plans.

MATERIALS
For each team:
- Math Balance kit with 49 tiles

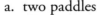 Instruct teams to calibrate their Math Balances before starting the test. Students should follow the steps exactly, and fill in either "safe" or "sunk" following each step of the loading plan. The canoe is "safe" if it remains balanced. It is "sunk" if either lever arm sinks to a height from the tabletop of less than 16 centimeters.

ASSESSMENT

Use the Rubric for Test, Communicate, and Redesign Steps on page 132 to assess how well students followed testing procedures.

INTERESTING INFO

There are many minimum safety precautions that must be taken before and during a canoeing trip. The canoe should have the following:

a. two paddles
b. extra rope for emergencies
c. a whistle to signal for help
d. a first-aid kit
e. a repair kit, such as duct tape or sealant
f. a life jacket for each passenger

6. TEST YOUR SOLUTION: BALANCING ACT!

Use the seesaw model to test your plan by following the steps shown in your drawing on page 111, and using the stacks you assembled in step 5. First, calibrate the balance using the white sliders under the lever arms so that the distance between the bottom edge of the lever arm on either side is about 19.8 centimeters. After each loading step, if the arms do not appear level, measure the distance between the bottom-edge corner of the lower arm and the tabletop (see diagram below). As long as the distance measurement is higher than or equal to 16 centimeters, you are SAFE. If the distance measurement is less than 16 centimeters, your canoe has SUNK.

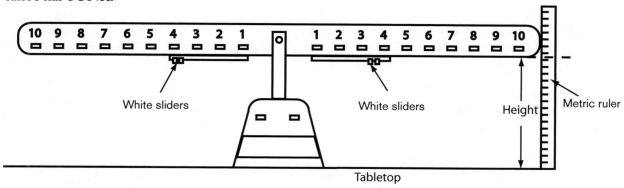

After each loading step, record whether your canoe is "SAFE" or "SUNK" in the table below.

STEP	CIRCLE ZONE	
Step 1	☺ SAFE	☹ SUNK
Step 2	☺ SAFE	☹ SUNK
Step 3	☺ SAFE	☹ SUNK
Step 4	☺ SAFE	☹ SUNK
Step 5	☺ SAFE	☹ SUNK
Step 6	☺ SAFE	☹ SUNK
Step 7	☺ SAFE	☹ SUNK
Step 8	☺ SAFE	☹ SUNK
Step 9	☺ SAFE	☹ SUNK
Step 10	☺ SAFE	☹ SUNK
Step 11	☺ SAFE	☹ SUNK
Step 12	☺ SAFE	☹ SUNK

You may use the Rubric for Test, Communicate, and Redesign Steps provided by your teacher to assess your work on the next few pages.

7. COMMUNICATE YOUR SOLUTION: BALANCING ACT!

OBJECTIVE: Students will answer questions to reflect on their designs, present their designs to other teams, and comment on one another's designs.

1. **TEAMS** Teams should discuss and answer questions 1 and 2. Tell them to select a representative to talk about their design, and share the test outcomes and reflection with the rest of the class.

2. **CLASS** Allow teams to take turns presenting.

7. COMMUNICATE YOUR SOLUTION: BALANCING ACT!

1. Based on the results of your test, do you think your loading plan was a success? If yes, why? If no, why not?

    _____

    _____

    _____

    _____

    _____

2. Which, if any, of the steps in your loading plan caused the canoe to "sink"? Why do you think the canoe sank? Is there a mathematical explanation for what happened?

    _____

    _____

    _____

    _____

    _____

8. REDESIGN AS NEEDED: BALANCING ACT!

OBJECTIVE: Students will answer questions to consider how they can redesign their canoe loading plans.

1.   TEAMS   Instruct teams to use what they learned from other teams' designs—both successes and failures—to improve their plan by answering questions 1 and 2.

2.   CLASS   Wrap up the activity by asking the class:
   • How did you use math to solve this loading problem?
   • What other knowledge was useful for designing a successful loading plan?

INTERESTING INFO
Engineers use a lot of math in their work. Basic mathematical concepts and algebraic equations allow engineers to formulate theories and perform calculations. Using statistics and probability, engineers can test their hypotheses by analyzing data where samples tested may contain variability or where experimental error is likely to occur. Engineers use data analysis, such as filtering and coding information, to describe, summarize, and compare the data with their initial hypotheses. Finally, engineers use modeling and simulations to predict the behavior and performance of their designs before they are actually built.

REDESIGN

8. REDESIGN AS NEEDED: BALANCING ACT!

1. Based on the results of your test, how would you change your design plan?

    _____

    _____

    _____

    _____

    _____

2. Explain, mathematically, why these changes would improve your loading plan.

    _____

    _____

    _____

    _____

    _____

Teacher Page

INDIVIDUAL SELF-ASSESSMENT RUBRIC: BALANCING ACT!

OBJECTIVE: Students will use a rubric to individually assess their involvement and work in this design challenge.

ASSESSMENT

Assign this reflection exercise as homework. You can write your comments on the lines below the self-assessment, and/or use this in conjunction with the Student Participation Rubric on page 133.

STUDENT PAGE

INDIVIDUAL SELF-ASSESSMENT RUBRIC: BALANCING ACT!

Use this rubric to reflect on how well you met behavior and work expectations during this activity. Check the box next to each expectation that you successfully met.

LEVEL 1	LEVEL 2	LEVEL 3	LEVEL 4	BONUS POINTS
Beginning to meet expectations	Meets some expectations	Meets expectations	Exceeds expectations	
☐ I was willing to work in a group setting.	☐ I met all of the Level 1 requirements.	☐ I met all of the Level 2 requirements.	☐ I met all of the Level 3 requirements.	☐ I helped resolve conflicts on my team.
☐ I was respectful and friendly to my teammates.	☐ I recorded the most essential comments from other group members.	☐ I made sure that my team was on track and doing the tasks for each activity.	☐ I helped my teammates understand the things that they did not understand.	☐ I responded well to criticism.
☐ I listened to my teammates and let them fully voice their opinions.	☐ I read all instructions.	☐ I listened to what my teammates had to say and asked for their opinions throughout the activity.	☐ I was always focused and on task: I didn't need to be reminded to do things; I knew what to do next.	☐ I encouraged everyone on my team to participate.
☐ I made sure we had the materials we needed and knew the tasks that needed to be done.	☐ I wrote down everything that was required for the activity.	☐ I actively gave feedback (by speaking and/or writing) to my team and other teams.	☐ I was able to explain to the class what we learned and did in the activity.	☐ I encouraged my team to persevere when my teammates faced difficulties and wanted to give up.
	☐ I listened to instructions in class and was able to stay on track.	☐ I completed all the assigned homework.		☐ I took advice and recommendations from the teacher about improving team performance and used feedback in team activities.
	☐ I asked questions when I didn't understand something.	☐ I was able to work on my own when the teacher couldn't help me right away.		☐ I worked with my team outside of the classroom to ensure that we could work well in the classroom.
		☐ I completed all the specified tasks for the activity.		

Approximate your level based on the number of checked boxes: _____   Bonus points: _____

Teacher comments: _____

Stranded!

TEAM EVALUATION: BALANCING ACT!

OBJECTIVE: Students will evaluate and discuss how well they worked in teams.

ASSESSMENT

1. **INDIVIDUALS** Assign this reflection exercise as homework or during quiet classroom time.

2. **TEAMS** Instruct students to share their team evaluation reflections with one another.

3. **CLASS** Point out any good examples of teamwork.

TEAM EVALUATION: BALANCING ACT!

How well did your team work together to complete the balancing design challenge? Reflect on your teamwork experience by completing this evaluation and sharing your thoughts with your team. Celebrate your successes!

RATE YOUR TEAMWORK. On a scale of 0–3, how well did your team do? 3 is **excellent;** 0 is **very poor.** Explain how you came up with that rating. Was it the same, better, or worse than the last activity?

LIST THINGS THAT WORKED WELL. Example: We got to our tasks right away and stayed on track.

LIST THINGS THAT DID NOT WORK WELL. Example: We argued a lot and did not come to a decision that everyone could agree on.

HOW CAN YOU IMPROVE TEAMWORK? Make the action steps concrete. For example: We need to learn how to make decisions better. Therefore, I will listen and respond without raising my voice.

Resources & Appendices

EDP: ENGINEERING DESIGN PROCESS

Engineers all over the world have one thing in common. They use the engineering design process (EDP) to solve problems. These problems can be as complicated as building a state-of-the-art computer or as simple as making a warm jacket. In both cases, engineers use the EDP to help solve the problems. Although engineers may not strictly follow every step of the EDP in the same order all the time, the EDP serves as a tool that helps to guide engineers in their thinking process and approach to a problem. Below is a brief outline of each step.

DEFINE	The first step is to **define** the problem. In doing so, remember to ask questions! What is the problem? What do I want to do? What specifications should my solution meet to successfully solve the problem (also called "criteria")? What factors may limit possible solutions to this problem (also called "constraints")?
RESEARCH	The next step is to conduct **research** on what can be done to solve the problem. What are the possible solutions? What have others already done? Use the Internet and the library to conduct investigations and talk to experts to explore possible solutions.
BRAINSTORM	**Brainstorm** ideas and be creative! Think about possible solutions in both two and three dimensions. Let your imagination run wild. Talk with your teacher and fellow classmates.

4 CHOOSE	**Choose** the best solution that meets all the criteria and constraints. Any diagrams or sketches will be helpful for later engineering design steps. Make a list of all the materials the project will need.
5 BUILD	Use your diagrams and list of materials as a guide to **build** a model or prototype of your solution.
6 TEST	**Test** and evaluate your prototype. How well does it work? Does it satisfy the engineering criteria and constraints?
7 COMMUNICATE	**Communicate** with your peers about your prototype. Why did you choose this design? Does it work as intended? If not, what could be fixed? What were the trade-offs in your design?
8 REDESIGN	Based on information gathered in the testing and communication steps, **redesign** your prototype. Keep in mind what you learned from one another in the communication step. Improvements can always be made!

MATH AND ENGINEERING CONCEPTS

In these three *Stranded!* activities, you will integrate engineering with math to solve problems and design prototypes. These activities should help you to:

- solve real-world problems involving distance, rate, and time

- solve problems involving proportions and scaling, and build scale models

- use formulas to calculate surface area for common geometric shapes (such as cylinders, rectangular prisms, and cones)

- understand and apply the engineering design process to solve problems

IMPORTANT VOCABULARY TERMS

ENGINEERING
the application of math and science to practical ends, such as design or manufacturing

ENGINEERING CONSTRAINTS
limiting factors to consider when designing a model

ENGINEERING CRITERIA
specifications met by a successful solution

PROTOTYPE
a test model that serves as a basis or standard for later stages

SCALE
a proportion used as a constant relationship between the dimensions of a model and the real object it represents

SCALE MODEL
an object that has been built to represent another, usually larger, object (The model is the same shape and has the same proportions, but is not the same actual size.)

RUBRIC FOR ENGINEERING DRAWINGS (STEP 4): A STORM IS APPROACHING!

	EXPERT (4)	COMPETENT (3)	BEGINNER (2)	NOVICE (1)
Quality of idea	☐ Selects a design that addresses the problem ☐ Uses materials efficiently and with purpose ☐ Chooses design as a team through thoughtful deliberation ☐ Is able to express rationale for each part of the design	☐ Selects a design that addresses most of the problem ☐ Uses material with some efficiency and with purpose ☐ Chooses design as a team after some deliberation ☐ Is able to express rationale for most parts of the design	☐ Selects a design that somewhat addresses the problem ☐ Uses materials with some purpose ☐ Chooses design as a team after a little discussion ☐ Is able to express rationale for some parts of the design	☐ Selects a design that does not address the problem ☐ Does not use materials with purpose ☐ Chooses design hastily without much discussion ☐ Is not able to express rationale for design
Communication	☐ Labels all dimensions ☐ Uses appropriate units ☐ Labels all the materials used in the design ☐ Indicates the scale ☐ Accurately draws the design to scale	☐ Labels most dimensions ☐ Uses appropriate units ☐ Labels most of the materials used in the design ☐ Indicates the scale ☐ Draws the design to scale, with minor errors	☐ Labels some dimensions. ☐ Uses somewhat appropriate units ☐ Labels some of the materials used in the design ☐ Indicates a scale, but it is inappropriate or incorrect ☐ Attempts to draw the design to scale	☐ Does not label dimensions ☐ Units are not included or are inappropriate. ☐ Does not label materials used in the design ☐ Does not indicate a scale ☐ Does not draw the design to scale

RUBRIC FOR ENGINEERING DRAWINGS (STEP 4): WE NEED WATER!

	EXPERT (4)	COMPETENT (3)	BEGINNER (2)	NOVICE (1)
Quality of idea	☐ Selects a design that addresses the problem ☐ Uses materials efficiently and with purpose ☐ Chooses design as a team through thoughtful deliberation ☐ Is able to express rationale for each part of the design	☐ Selects a design that addresses most of the problem ☐ Uses material with some efficiency and with purpose ☐ Chooses design as a team after some deliberation ☐ Is able to express rationale for most parts of the design	☐ Selects a design that somewhat addresses the problem ☐ Uses materials with some purpose ☐ Chooses design as a team after a little discussion ☐ Is able to express rationale for some parts of the design	☐ Selects a design that does not address the problem ☐ Does not use materials with purpose ☐ Chooses design hastily without much discussion ☐ Is not able to express rationale for design
Communication	☐ Labels all dimensions ☐ Uses appropriate units ☐ Labels all the materials used in the design	☐ Labels most dimensions ☐ Uses appropriate units ☐ Labels most of the materials used in the design	☐ Labels some dimensions. ☐ Uses somewhat appropriate units ☐ Labels some of the materials used in the design	☐ Does not label dimensions ☐ Units are not included or are inappropriate. ☐ Does not label materials used in the design

RUBRIC FOR CANOE-LOADING PLAN (STEP 4): BALANCING ACT!

	EXPERT (4)	COMPETENT (3)	BEGINNER (2)	NOVICE (1)
Quality of idea	☐ Includes all people and items in plan ☐ Chooses design as a team through thoughtful deliberation ☐ Is able to express rationale for each part of the design	☐ Includes most of the people and items in plan ☐ Chooses design as a team after some deliberation ☐ Is able to express rationale for most parts of the design	☐ Includes some of the people and items in plan ☐ Chooses design as a team after a little discussion ☐ Is able to express rationale for some parts of the design	☐ Includes a few or none of the people and items in plan ☐ Chooses design hastily without much discussion ☐ Is not able to express rationale for design
Communication	☐ Labels all items and people ☐ Numbers all the items and people in order of loading ☐ Shows correct mathematical justification for plan	☐ Labels most of the items and people ☐ Numbers most of the items and people in order of loading ☐ Shows a mostly correct mathematical justification for the plan (minor errors in calculation or sums off by 1 or 2)	☐ Labels some items and people ☐ Numbers some of the items and people in order of loading ☐ Attempts to show some mathematical justification for the plan, but with major errors	☐ Does not label items and people ☐ Numbers a few or none of the items and people in order of loading ☐ Does not show a mathematical justification for the plan

RUBRIC FOR PROTOTYPE/MODEL (STEP 5): A STORM IS APPROACHING!, WE NEED WATER!, AND BALANCING ACT!

	EXPERT (4)	COMPETENT (3)	BEGINNER (2)	NOVICE (1)
Completeness	☐ Builds a model that meets all criteria and constraints ☐ Follows the design sketch ☐ Follows cleanup procedures	☐ Builds a model that addresses most of the criteria and constraints ☐ Follows most of the design sketch ☐ Follows cleanup procedures	☐ Builds a model that addresses some of the criteria and constraints ☐ Follows some of the design sketch ☐ Partially follows cleanup procedures	☐ Builds an incomplete model ☐ Does not follow the design sketch ☐ Does not follow cleanup procedures
Craftsmanship	☐ Takes care in constructing model; is adept with tools and resources, making continual adjustments to adjust the model/prototype ☐ Demonstrates persistence with minor problems	☐ Uses tools and resources with little or no guidance ☐ Refines model to enhance appearance and capabilities	☐ Uses tools and resources with some guidance; may have difficulty selecting the appropriate resource ☐ Refines work, but may prefer to leave model as first produced	☐ Needs guidance in order to use resources safely and appropriately ☐ Model/prototype is crude, with little or no refinements made

RUBRIC FOR TEST, COMMUNICATE, AND REDESIGN STEPS: A STORM IS APPROACHING!, WE NEED WATER!, AND BALANCING ACT!

	EXPERT (4)	COMPETENT (3)	BEGINNER (2)	NOVICE (1)
Completeness	☐ Carefully follows the testing procedures and documents all testing results	☐ Follows the testing procedures and documents most of the testing results	☐ Follows some of the testing procedures and documents some of the testing results	☐ Does not follow the testing procedures and does not document the testing results
Model performance	☐ The model fully meets the design constraints and criteria.	☐ The model meets most of the design constraints and criteria.	☐ The model meets some design constraints and criteria completely but ignores others.	☐ The model fails to meet design criteria and constraints.
Quality of reflection	☐ Specific improvement ideas are generated and documented.	☐ Some general improvement ideas are generated and documented.	☐ The need for improvements is recognized and some ideas are generated, but documentation is not complete.	☐ Little interest is taken in improving the prototype or model, despite problems detected during testing. There is no evidence of inclination or ability to generate refinement solutions.

Name
STUDENT PAGE

STUDENT PARTICIPATION RUBRIC: A STORM IS APPROACHING!, WE NEED WATER!, AND BALANCING ACT!

	0 POINTS	1 POINT	2 POINTS	3 POINTS	SCORE
CONTENT CONTRIBUTION					
Sharing information	Discussed very little information related to the topic	Discussed some basic information; most related to the topic	Discussed a great deal of information; all related to the topic	Discussed a great deal of information showing in-depth analysis and thinking skills	
Creativity	Did not contribute any new ideas	Contributed some new ideas	Contributed many new ideas	Contributed a great deal of new ideas	
RESPONSIBILITY					
Completion of assigned duties	Did not perform any assigned duties	Performed very few assigned duties	Performed nearly all assigned duties at the level of expectation	Performed all assigned duties; did extra duties	
Attendance	Was never present or was always a negative influence when present	Attended some group meetings; absence(s) hurt the group's progress	Attended most group meetings; absence(s) did not affect group's progress or made up work	Attended all focus group meetings	
Staying on task	Not productive during group meetings; often distracted the team	Productive some of the time; needed reminders to stay on task	Productive most of the time; rarely needed reminders to stay on task	Used all of the focus group time effectively; productive at all times	
TEAMWORK					
Cooperating with teammates	Was rarely talking or always talking; usually argued with teammates	Usually did most of the talking and rarely allowed others to speak; sometimes argued	Listened, but sometimes talked too much; rarely argued	Listened and spoke a fair amount; never argued with teammates	
Making fair decisions	Always needed to have things his or her way; easily upset	Usually wanted to have things his or her way or often sided with friends instead of considering all views	Usually considered all views	Always helped team reach a fair decision	
Leadership	Never took lead; needed to be assigned duties	Took a lead at least once; volunteered for duty	Took a lead more than once; volunteered for duties and helped others	Played an essential role in organizing the group; frequently took lead; always helped others	

Total score: _____ / 24

Name: _____

Teacher: _____

STUDENT WORK SAMPLE 1
A STORM IS APPROACHING! STEP 4: CHOOSE THE BEST SOLUTION

Use the rubric provided by your teacher to assess the following student work sample. Write a brief explanation for the grade you assign and how the work can be improved.

1. Grade: _____

2. Reasons for grade: _____

    _____

    _____

    _____

    _____

3. How work can be improved: _____

    _____

    _____

    _____

    _____

STUDENT WORK SAMPLE 2
A STORM IS APPROACHING! STEP 4: CHOOSE THE BEST SOLUTION

Use the rubric provided by your teacher to assess the following student work sample. Write a brief explanation for the grade you assign and how the work can be improved.

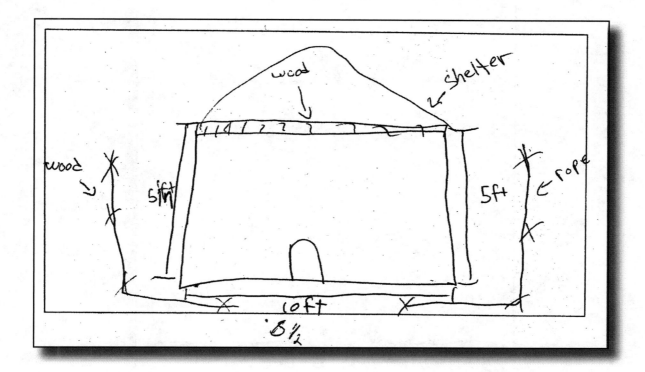

1. Grade: _____

2. Reasons for grade: _____

   _____

   _____

   _____

   _____

3. How work can be improved: _____

   _____

   _____

   _____

   _____

STUDENT WORK SAMPLE 3
A STORM IS APPROACHING! STEP 4: CHOOSE THE BEST SOLUTION

Use the rubric provided by your teacher to assess the following student work sample. Write a brief explanation for the grade you assign and how the work can be improved.

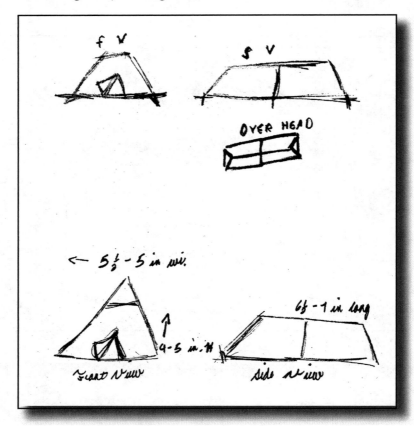

1. Grade: _____

2. Reasons for grade: _____

    _____

    _____

    _____

    _____

3. How work can be improved: _____

    _____

    _____

    _____

    _____

STUDENT WORK SAMPLE 4

A STORM IS APPROACHING! STEP 4: CHOOSE THE BEST SOLUTION

Use the rubric provided by your teacher to assess the following student work sample. Write a brief explanation for the grade you assign and how the work can be improved.

1. Grade: _____

2. Reasons for grade: _____
    _____
    _____
    _____
    _____

3. How work can be improved: _____
    _____
    _____
    _____
    _____

STUDENT WORK SAMPLE 5
A STORM IS APPROACHING! STEP 4: CHOOSE THE BEST SOLUTION

Use the rubric provided by your teacher to assess the following student work sample. Write a brief explanation for the grade you assign and how the work can be improved.

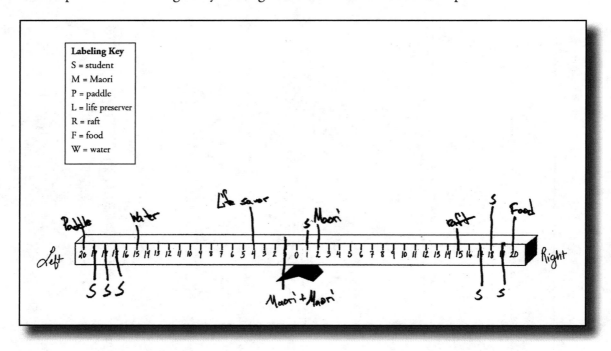

1. Grade: _____

2. Reasons for grade: _____
   _____
   _____
   _____
   _____

3. How work can be improved: _____
   _____
   _____
   _____
   _____
   _____

STUDENT WORK SAMPLE 6

A STORM IS APPROACHING! STEP 4: CHOOSE THE BEST SOLUTION

Use the rubric provided by your teacher to assess the following student work sample. Write a brief explanation for the grade you assign and how the work can be improved.

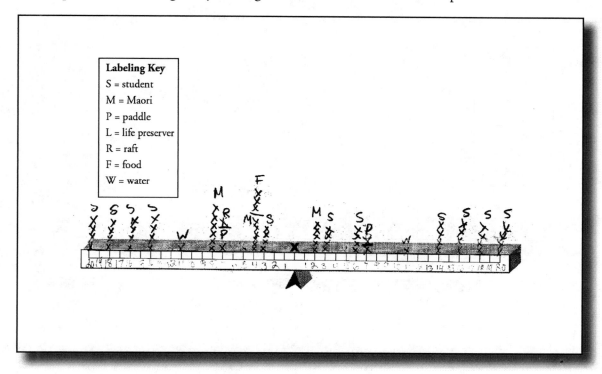

Labeling Key
S = student
M = Maori
P = paddle
L = life preserver
R = raft
F = food
W = water

1. Grade: _____

2. Reasons for grade: _____

   _____

   _____

   _____

   _____

3. How work can be improved: _____

   _____

   _____

   _____

   _____

   _____

STUDENT WORK SAMPLE 7
A STORM IS APPROACHING! STEP 4: CHOOSE THE BEST SOLUTION

Use the rubric provided by your teacher to assess the following student work sample.
Write a brief explanation for the grade you assign and how the work can be improved.

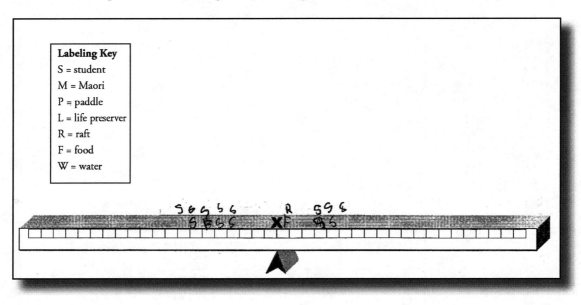

Labeling Key
S = student
M = Maori
P = paddle
L = life preserver
R = raft
F = food
W = water

1.   Grade: _____

2.   Reasons for grade: _____

_____

_____

_____

_____

_____

3.   How work can be improved: _____

_____

_____

_____

_____

_____

STUDENT WORK SAMPLE 8
A STORM IS APPROACHING! STEP 4: CHOOSE THE BEST SOLUTION

Use the rubric provided by your teacher to assess the following student work sample. Write a brief explanation for the grade you assign and how the work can be improved.

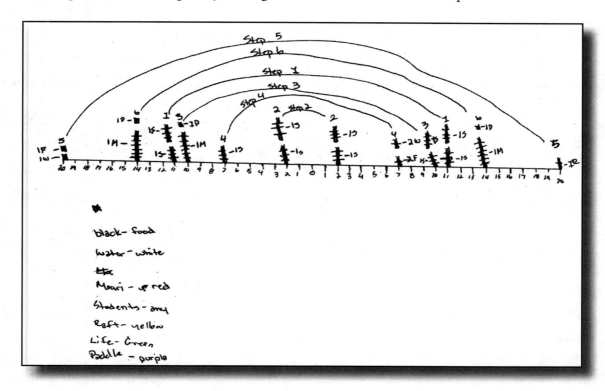

1. Grade: _____

2. Reasons for grade: _____
   _____
   _____
   _____
   _____

3. How work can be improved: _____
   _____
   _____
   _____
   _____
   _____

BOX *A*

1. Copy onto cardstock.
2. Cut out net, fold, assemble, and tape.

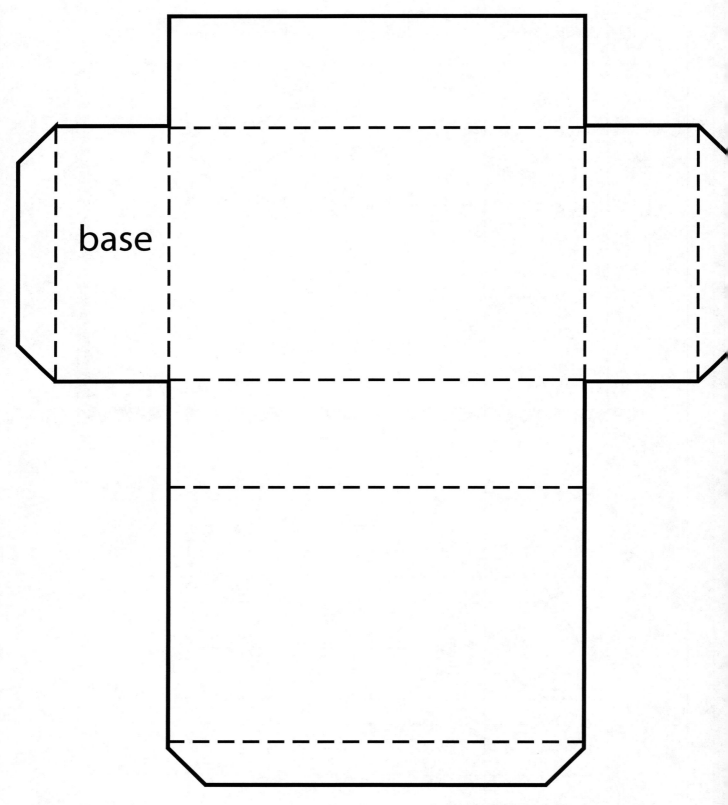

BOX *B*

1. Photocopy and cut out rectangles *X, Y,* and *Z.*

2. Trace the rectangles onto poster board according to the net diagram on the right.

3. Cut out the net, fold, and tape.

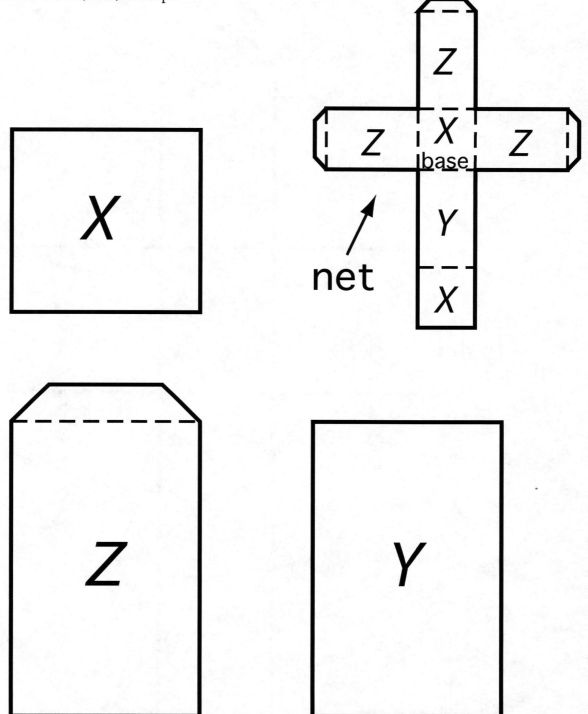

BOX *C*

1. Copy onto cardstock.
2. Cut out net, fold, assemble, and tape.

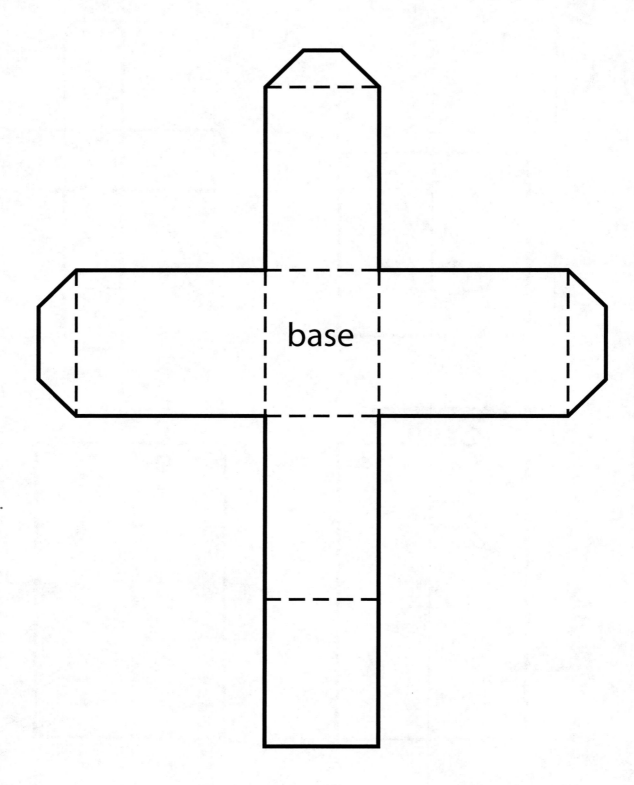

BOX *D*

1. Copy onto cardstock.
2. Cut out net, fold, assemble, and tape.

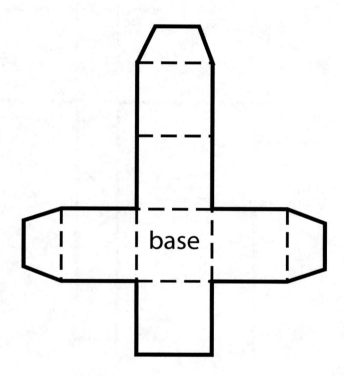

BOX *E*

1.   Photocopy and cut out rectangles *X, Y, Z,* and Z_F on the following pages (include the flap).

2.   Trace the rectangles onto poster board according to the net diagram on the right.

3.   Cut out the net, fold, and tape.

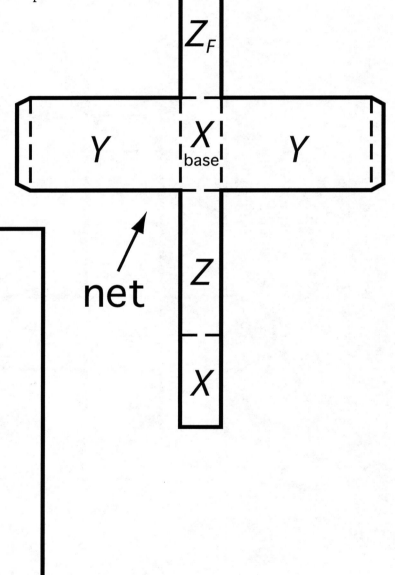

net

BOX *E*

BOX *E*

Z

Z_F

BOX *F*

1. Photocopy and cut out rectangles *X, Y,* and Y_F on the following pages (include the flap).

2. Trace the rectangles onto poster board according to the net diagram on the right.

3. Cut out the net, fold, and tape.

net

$$
\begin{array}{c}
Y_F \\
Y_F \quad \begin{array}{|c|} X \\ \text{base} \end{array} \quad Y_F \\
Y \\
X
\end{array}
$$

X

BOX *F*

Y

BOX F

$$Y_F$$

BOX *G*

1. Photocopy and cut out rectangles *X*, *Y*, *Z*, and Z_F on the following pages (include the flap).

2. Trace the rectangles onto poster board according to the net diagram on the right.

3. Cut out the net, fold, and tape.

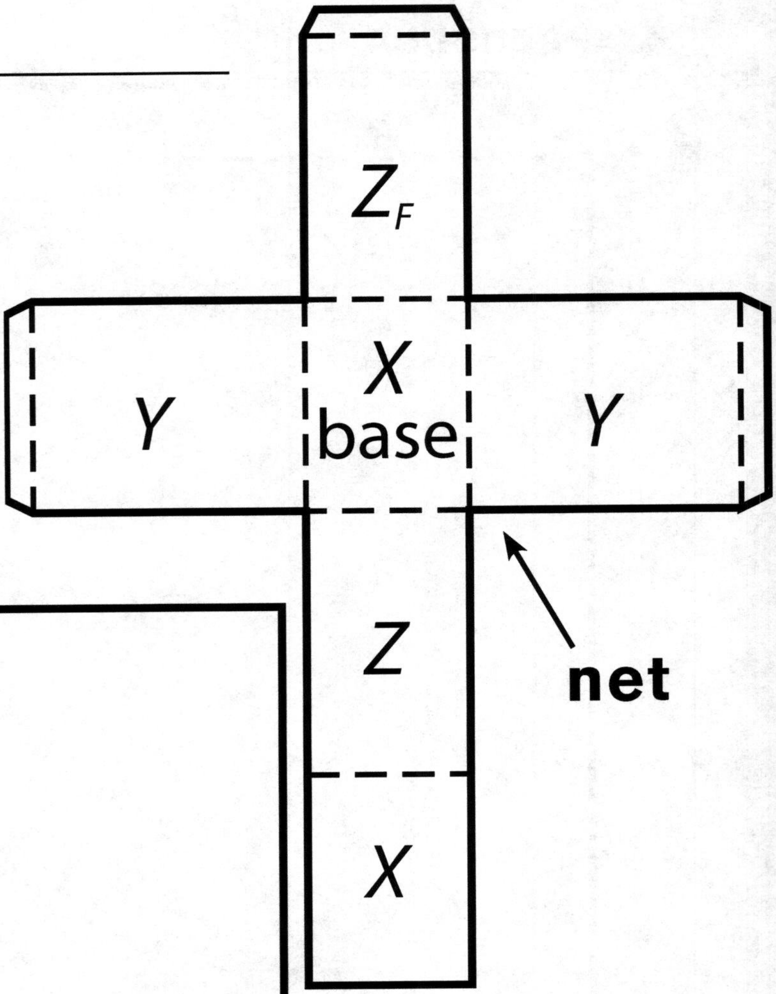

Z_F

Y   *X* base   *Y*

Z

X

net

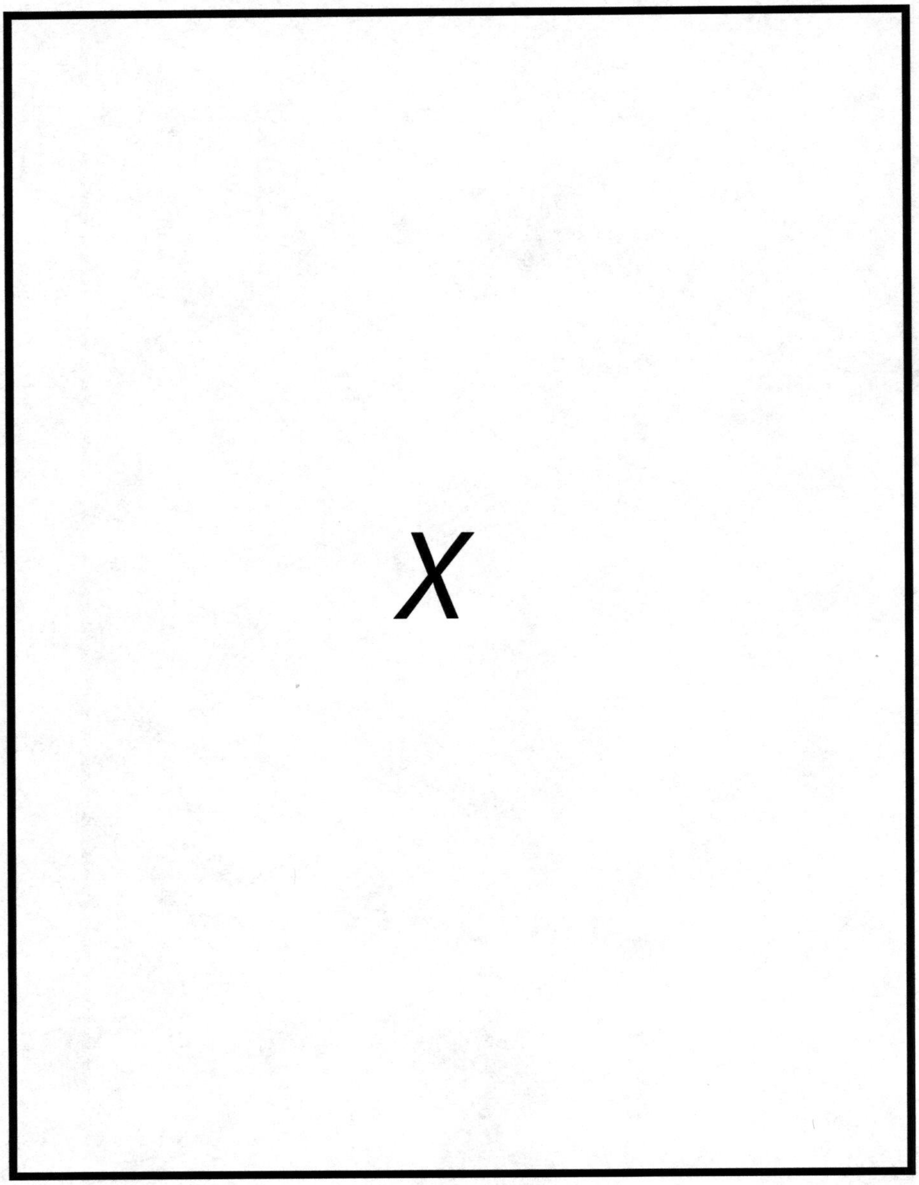

X

BOX *G*

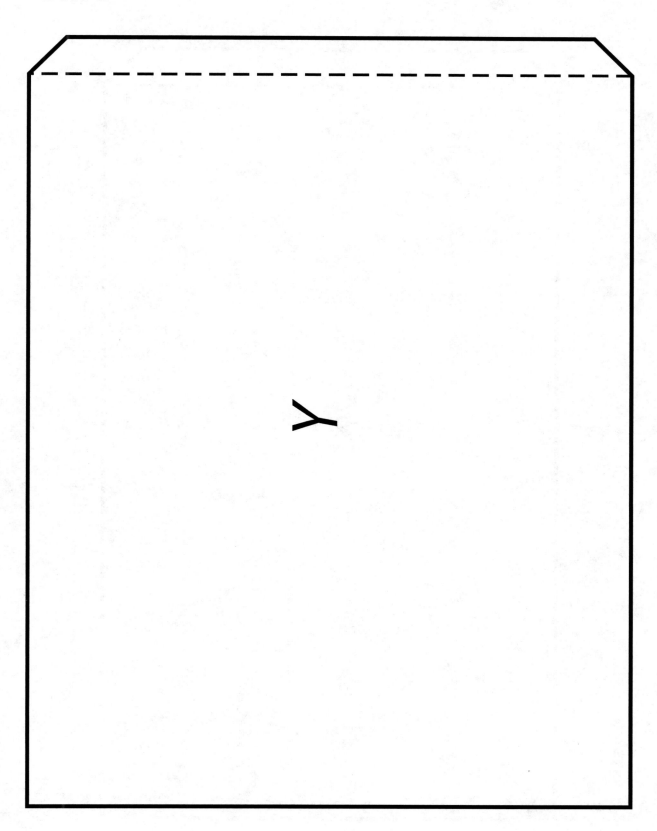

BOX *G*

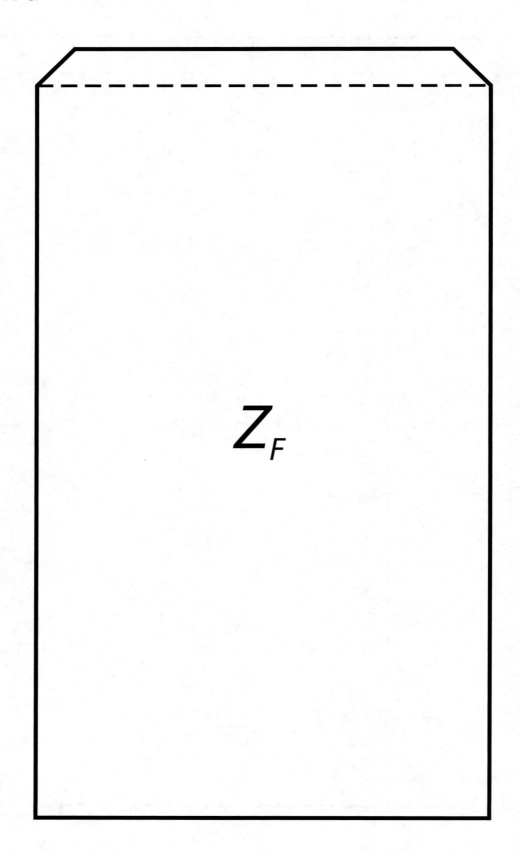

BOX *G*

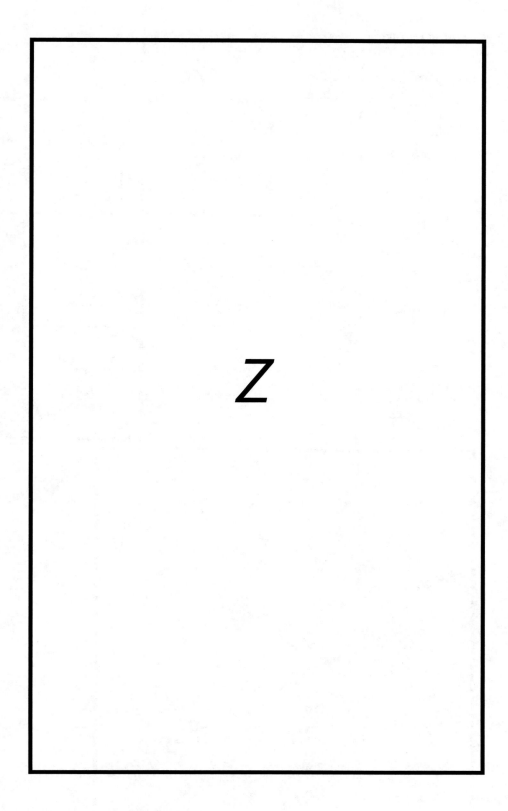

BOX *H*

1. Photocopy and cut out rectangles *X* and X_F on the following pages (include the flap).

2. Trace the rectangles onto poster board according to the net diagram on the right.

3. Cut out the net, fold, and tape.

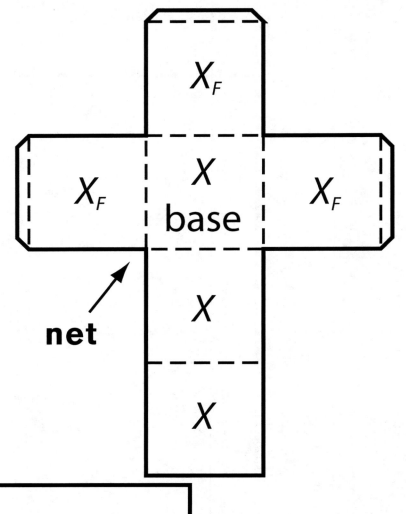

net

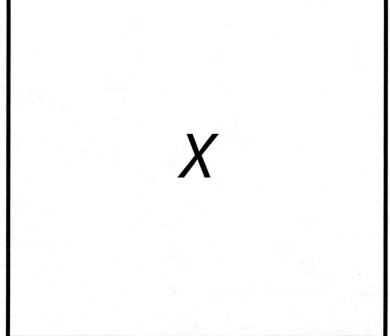

BOX *H*

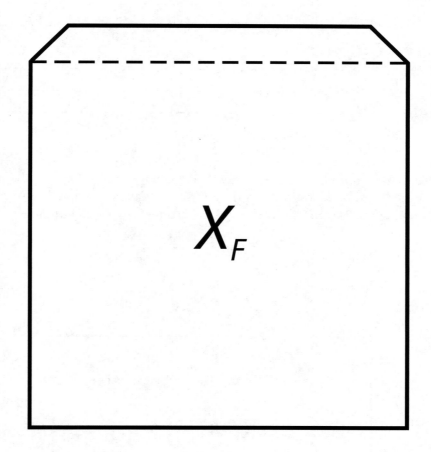

CYLINDERS (NET-MAKING DIRECTIONS)

1.   Measure and draw two of each rectangle on poster board according to the dimensions listed in the table below. Label each rectangle with its letter. Leave space above the dimension labeled "width" because that's where you'll trace the bottom of each cylinder.

Can	Height (cm)	Width (cm)
A	24.3	9.4
B	13.8	15.7
C	9.0	22.0
D	5.0	31.4
E	2.5	40.8
F	1.3	47.1

2.   Make two photocopy sets of circles A–F (on pages 160–162), cut them out, and glue the circle templates onto the corresponding rectangles so that each circle's edge is touching the "width" dimension on the rectangle. See diagram below.

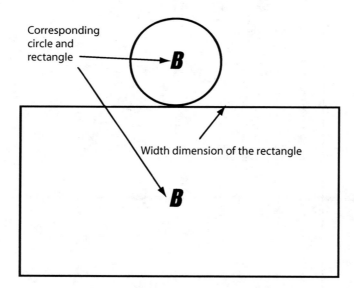

Corresponding circle and rectangle

Width dimension of the rectangle

3. Cut out one set of cylinder nets and tape the edges shut. Cut out the duplicate set of nets so that the circles and rectangles are separate pieces. Roll up the duplicate rectangles and place them in their corresponding cylinder. Velcro the bases of the cylinders and attach the duplicate circles to the bases. This will allow the students to remove the two parts of each cylinder and measure them to calculate the surface area without needing to actually unroll each cylinder. See diagrams below:

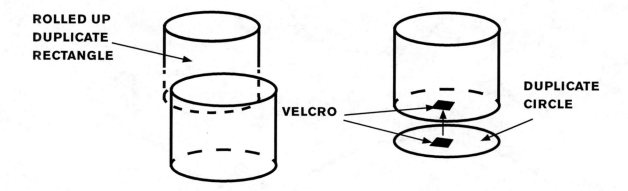

ROLLED UP DUPLICATE RECTANGLE

VELCRO

DUPLICATE CIRCLE

CYLINDERS (CIRCLE TEMPLATES *A–D*)

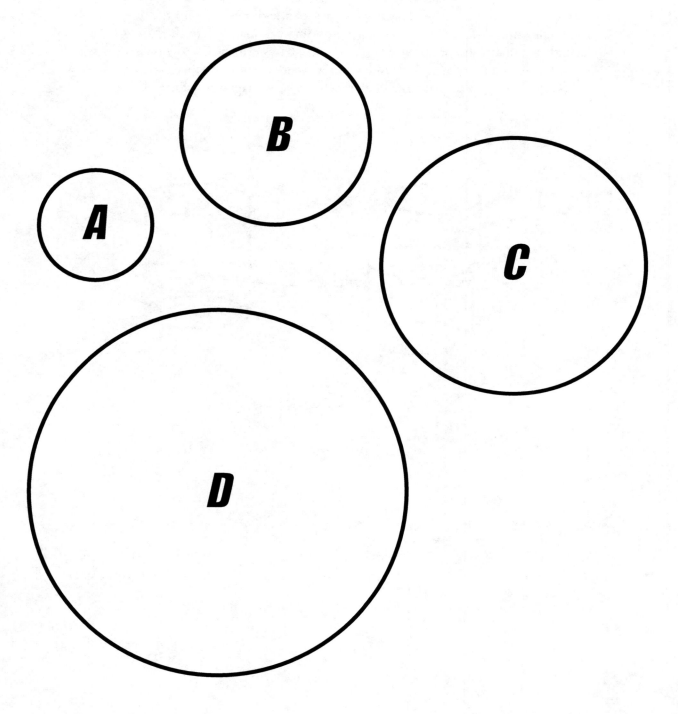

CYLINDERS (CIRCLE TEMPLATE *E*)

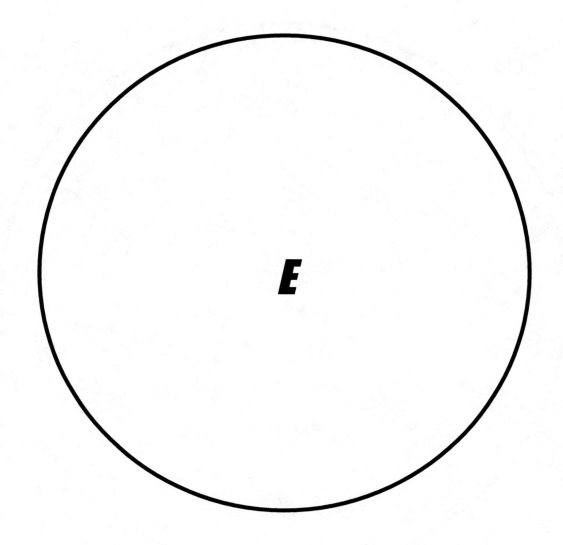

CYLINDERS (CIRCLE TEMPLATE *F*)

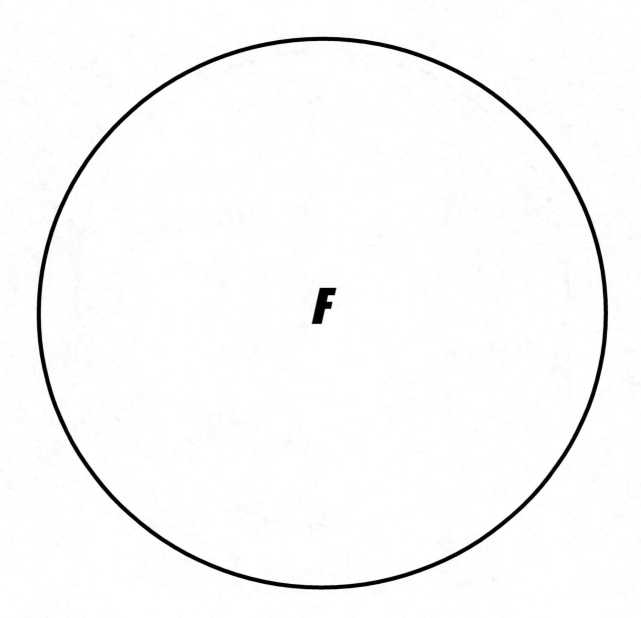

OTHER CYLINDER SETS (NET–MAKING DIRECTIONS)

1. Measure and draw rectangles on poster board according to the dimensions listed in the table below. Label each rectangle with its color and letter. Leave space above the dimension labeled "width," because that's where you'll trace the bottom of each cylinder.

Can	Height (cm)	Width (cm)
Orange-*A*	8.3	9.4
Orange-*B*	5.8	12.6
Orange-*C*	3.0	18.8
Orange-*D*	1.8	25.1
Orange-*E*	0.8	28.3

Can	Height (cm)	Width (cm)
Pink-*A*	11.5	9.4
Pink-*B*	6.1	15.8
Pink-*C*	3.5	22.0
Pink-*D*	1.8	28.3
Pink-*E*	1.2	31.4

Can	Height (cm)	Width (cm)
Yellow-*A*	11.0	12.6
Yellow-*B*	6.5	18.8
Yellow-*C*	4.0	25.1
Yellow-*D*	2.3	31.4
Yellow-*E*	1.6	34.5

2. Make a copy of the circle templates (pages 165–167), cut out the circles, and glue them onto the poster boards next to the corresponding rectangles so that each circle's edge is touching the "width" dimension on the rectangle. See diagram below.

O_B = ORANGE-*B*

P_B = PINK-*B*

Y_B = YELLOW-*B*

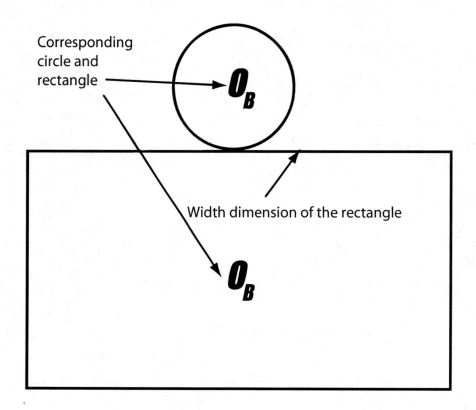

Corresponding circle and rectangle

Width dimension of the rectangle

3. Cut out each cylinder net and tape the edges together.

ORANGE CYLINDERS CIRCLE TEMPLATES

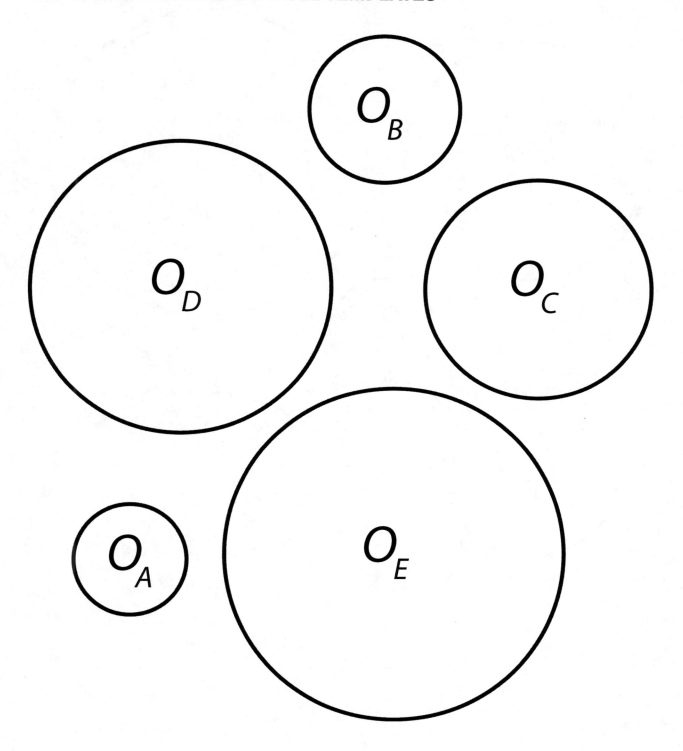

PINK CYLINDERS CIRCLE TEMPLATES

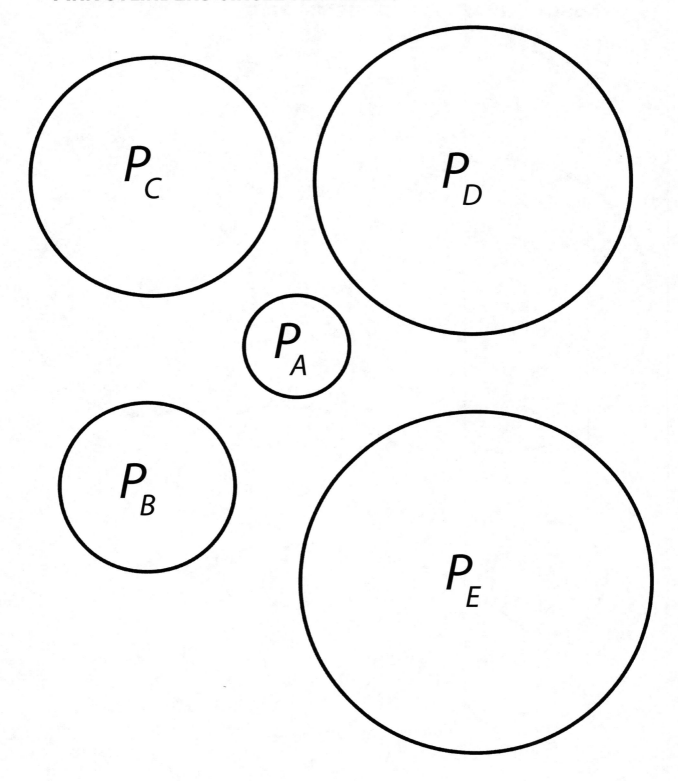

YELLOW CYLINDERS CIRCLE TEMPLATES

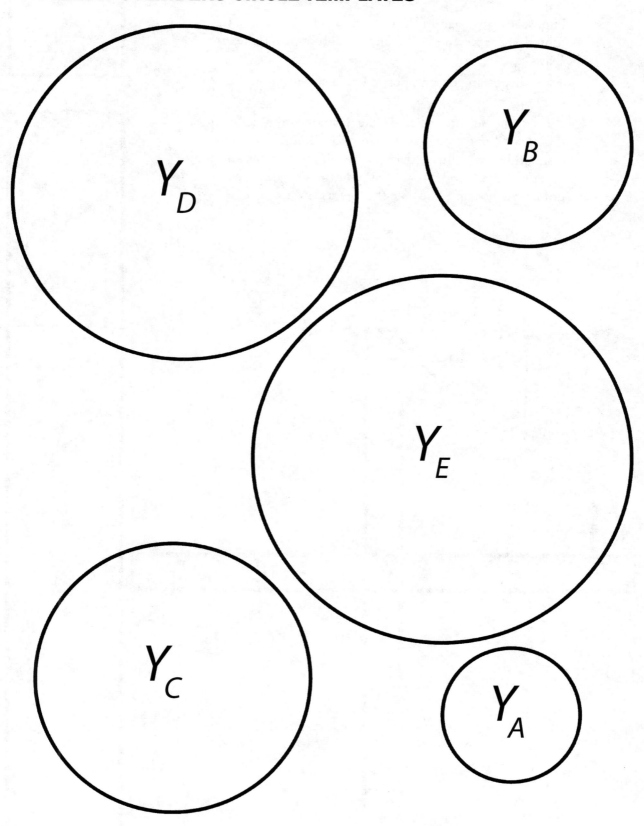

SQUARE PRISM *A*

1. Photocopy this page and cut out rectangle A_S and A_B. The subscript S represents "side" and the subscript B represents "base."

2. Trace the rectangles onto poster board according to the net diagram below.

3. Cut out the net you traced on the poster board, fold along the dotted lines, tape yarn or string to the outer sides (see diagram below), and tie strings with slipknots to keep the edges together. Students can easily untie the strings to unfold the box and calculate the prism's surface area.

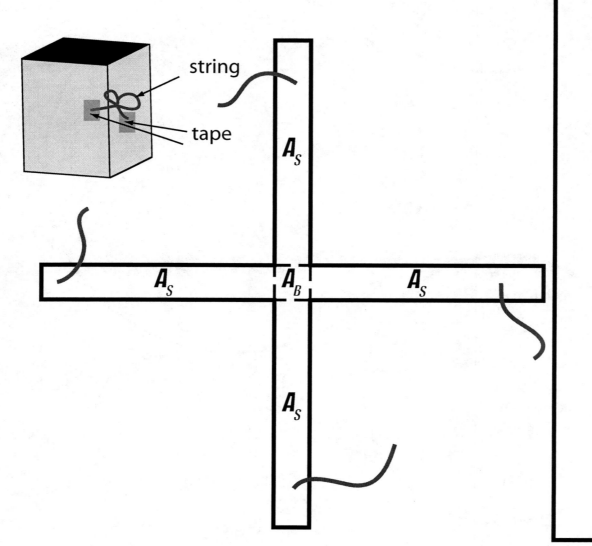

© Museum of Science (Boston), Wong, Brizuela

SQUARE PRISM *B*

1. Photocopy this page and cut out rectangle B_S and B_B. The subscript S represents "side" and the subscript B represents "base."

2. Trace the rectangles onto poster board according to the net diagram below.

3. Cut out the net you traced on the poster board, fold along the dotted lines, tape yarn or string to the outer sides (see diagram below), and tie strings with slipknots to keep the edges together. Students can easily untie the strings to unfold the box and calculate the prism's surface area.

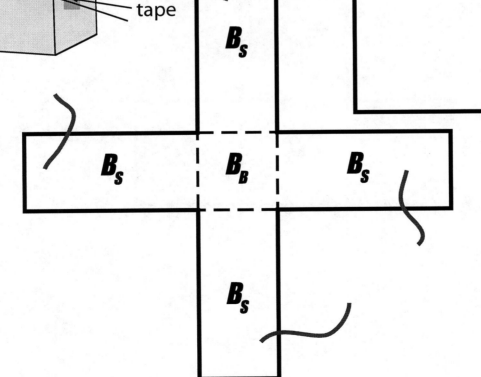

SQUARE PRISM C

1. Photocopy this page and cut out rectangle C_S and C_B. The subscript S represents "side" and the subscript B represents "base."

2. Trace the rectangles onto poster board according to the net diagram below.

3. Cut out the net you traced on the poster board, fold along the dotted lines, tape yarn or string to the outer sides (see diagram below), and tie strings with slipknots to keep the edges together. Students can easily untie the strings to unfold the box and calculate the prism's surface area.

C_B

C_S

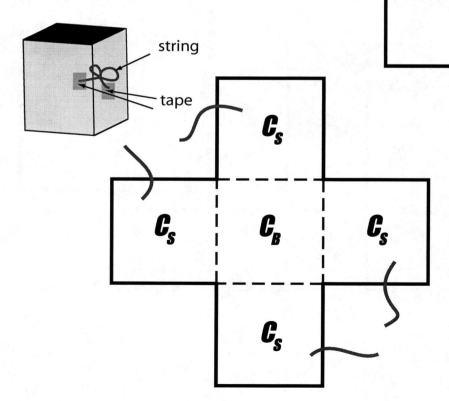

string

tape

C_S

C_S C_B C_S

C_S

SQUARE PRISM *D*

1. Photocopy this page onto cardstock.
2. Cut out the net, fold along the dotted lines, tape yarn or string to the outer sides (see diagram on the right), and tie the strings with slipknots to keep the edges together. Students can easily untie the strings to unfold the box and calculate the prism's surface area.

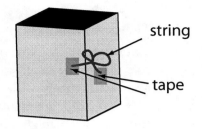

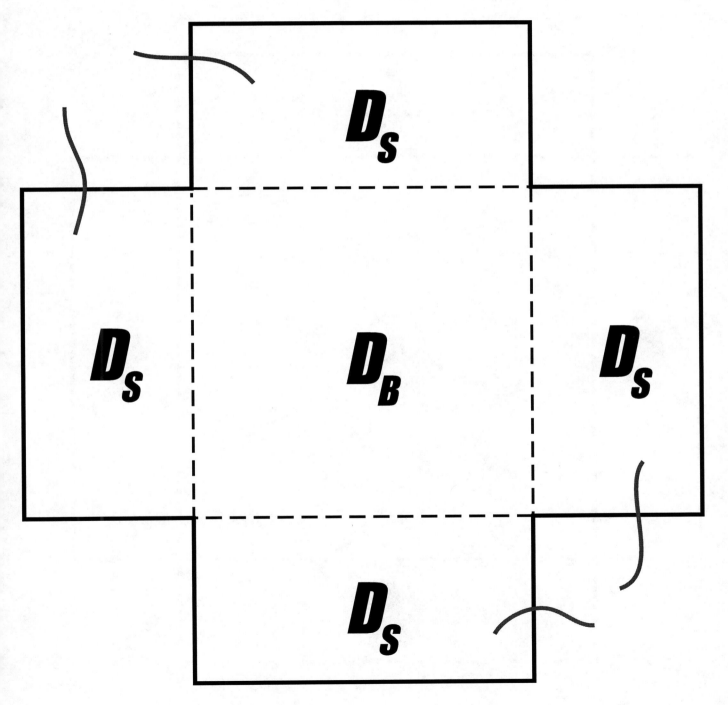

SQUARE PRISM *E*

1. Photocopy this page onto cardstock.
2. Cut out the net, fold along the dotted lines, tape yarn or string to the outer sides (see diagram on the right), and tie the strings with slipknots to keep the edges together. Students can easily untie the strings to unfold the box and calculate the prism's surface area.

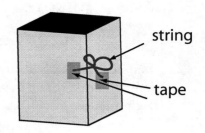

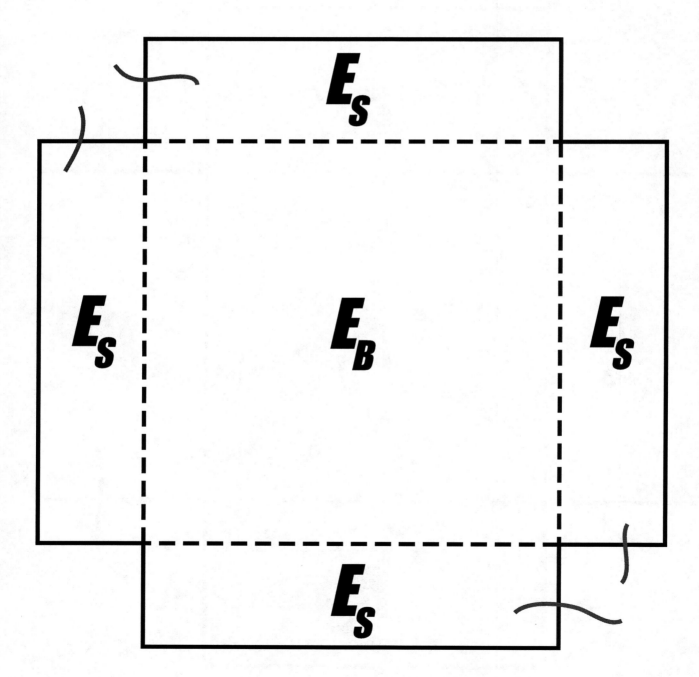

SQUARE PRISM *F*

1. Photocopy this page onto cardstock.
2. Cut out the net, fold along the dotted lines, tape yarn or string to the outer sides (see diagram on the right), and tie the strings with slipknots to keep the edges together. Students can easily untie the strings to unfold the box and calculate the prism's surface area.

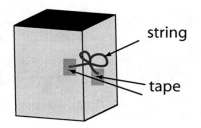

string

tape

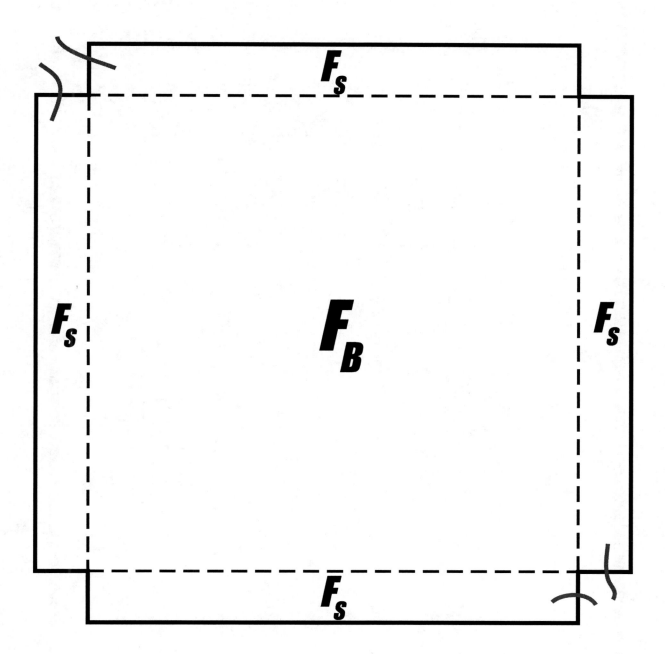

Answer Key

SPEED = DISTANCE/TIME ACTIVITY
1.   50 km
2.   6.67 km/hr
3.   1/3 hour or 20 minutes

USING A SCALE ACTIVITY
1.   1.5 cm
2.   30 cm × 100 cm
3.   12 cm × 12 cm × 12 cm = 1728 cm³ (equivalent to 1.728 liters)

WHERE ARE WE?
Step 1: 850 km/h × 9.5 hours = 8,075 km
Step 4: 15 km/h × 24 hours = 360 km

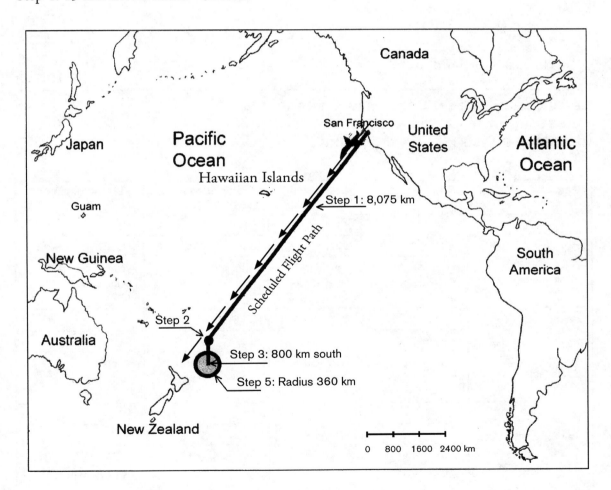

INTRODUCING THE ENGINEERING DESIGN PROCESS (EDP)

1. Answers will vary.

2. Step 1: Define the problem.
   Step 2: Research the problem.
   Step 3: Brainstorm possible solutions.
   Step 4: Choose the best solution.
   Step 5: Build a model or prototype.
   Step 6: Test your solution.
   Step 7: Communicate your solution.
   Step 8: Redesign as needed.

3. a. Step 2: Research
   b. Step 5: Build
   c. Step 7: Communicate
   d. Step 1: Define
   e. Step 4: Choose
   f. Step 8: Redesign
   g. Step 6: Test
   h. Step 3: Brainstorm

DESIGN CHALLENGE 1: A STORM IS APPROACHING!
2. RESEARCH THE PROBLEM
RESEARCH PHASE 1: SCALE MODELING

1.

Pair	Box	Height (cm)	Area of Base (cm²)	Volume (cm³)
Pair 1: A & E Scale factor = 2	A	11	21	231
	E	22	84	1848
	How many times bigger is E?	Height is 2 times bigger.	Area is $2^2 = 4$ times bigger.	Volume is $2^3 = 8$ times bigger.
Pair 2: B & F Scale factor = 3	B	8	25	200
	F	24	225	5400
	How many times bigger is F?	Height is 3 times bigger.	Area is $3^2 = 9$ times bigger.	Volume is $3^3 = 27$ times bigger.
Pair 3: C & G Scale factor = 4	C	5	12	60
	G	20	192	3840
	How many times bigger is G?	Height is 4 times bigger.	Area is $4^2 = 16$ times bigger.	Volume is $4^3 = 64$ times bigger.
Pair 4: D & H Scale factor = 5	D	2	4	8
	H	10	100	1000
	How many times bigger is H?	Height is 5 times bigger.	Area is $5^2 = 25$ times bigger.	Volume is $5^3 = 125$ times bigger.

ANSWER KEY

2.

For a scale factor = x

1-D	Length |------|	Scales to →	Length • x |--------------------|
2-D	Area	Scales to →	Area • x^2
3-D	Volume	Scales to →	Volume • x^3

Students should come to the conclusion that when we have two similar objects, the scale factor tells us how many times larger one-dimensional measurements of the larger object are compared to the smaller. For two-dimensional area measurements, the larger object is greater by the scale factor squared. For three-dimensional volume measurements, the larger object is greater by the scale factor cubed.

3. Students may choose any scale. However, it may be easiest if they choose the scale such that one craft stick represents one log. The solutions for this scaling choice are provided below.

   a. 4 centimeters = 1 meter
      1 log = 3 meters
      1 craft stick = 12 cm
      If 1 craft stick represents 1 log, then our scale must be 12 cm represents 3 meters, or
      4 cm represents 1 meter.

   b. This is an easy scale to work with because 1 stick in our model will represent 1 real log.

   c. Actual = 1 m • 1 m • 1 m
      Therefore, model = 4 cm • 4 cm • 4 cm = 64 cubic cm (cm³)

4. a. 20 sticks
   b. 10 cm • 16 cm
   c. 12 cm • 20 cm
   d. 24 cm
   e. 64 cubic cm (cm³)

DESIGN CHALLENGE 2: WE NEED WATER
2. RESEARCH THE PROBLEM
RESEARCH PHASE 1: FIND THE AREA OF THE PLANE SIDING

Some students may recognize that this figure is exactly 1/2 of a rectangle that is 55.88 cm • 71.12 cm.

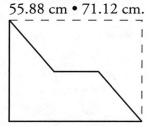

Therefore, the area of this figure is:

$$\frac{55.88 \cdot 71.12}{2} \approx 1987 \text{ cm}^2$$

Other ways to solve the problem include breaking up the shape into triangles, rectangles, or parallelograms. See examples below.

© Museum of Science (Boston), Wong, Brizuela

ANSWER KEY

RESEARCH PHASE 2: DOES SAME SURFACE AREA MEAN SAME VOLUME?

1. Answers will vary.

2. a. We found that two cylinders that are rolled using the same amount of material don't have the same volume. The shorter, wider one holds more than the taller, narrower cylinder.

   b. The volume of a cylinder, volume = $\pi \cdot radius^2 \cdot height$. Since the radius is squared (multiplied by itself) and the height is only multiplied once, the radius has a greater effect on the volume. The radius is related to volume quadratically, whereas the height is only related to the volume linearly.

RESEARCH PHASE 3: CYLINDERS

1.

Can	Radius (cm)	Height (cm)	Circumference (cm)	Surface Area (sq. cm or cm²)	Volume (mL)
A	1.5	24.3	9.4	≈236	172
B	2.5	13.8	15.7	≈236	271
C	3.5	9.0	22.0	≈236	346
D	5.0	5.0	31.4	≈236	393
E	6.5	2.5	40.8	≈236	332
F	7.5	1.3	47.1	≈236	230

2–3. Answers will vary.

4.

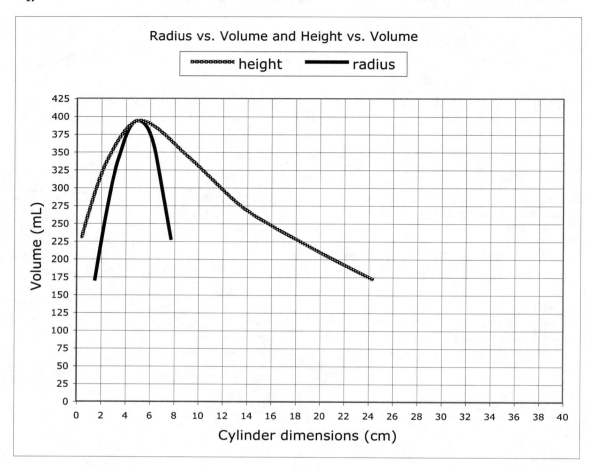

5. If surface area is kept constant, the height decreases as the radius increases.

6. If surface area is kept constant, the radius decreases as the height increases.

7. Cone D has the largest volume; radius and height are equal.

8. The volume will be greatest when radius equals height.

MORE CYLINDERS (OPTIONAL ACTIVITY)

Orange set: surface area ≈ 85 cm²

	A	B	C	D	E
Radius (cm)	1.5	2	3	4	4.5
Height (cm)	8.3	5.8	3	1.8	0.8
Volume (cm³)	59	73	85	90	51

Pink set: surface area ≈ 115 cm²

	A	B	C	D	E
Radius (cm)	1.5	2.5	3.5	4.5	5
Height (cm)	11.5	6.1	3.5	1.8	1.2
Volume (cm³)	81	120	135	115	94

ANSWER KEY

Yellow set: surface area ≈ 151 cm^2

	A	B	C	D	E
Radius (cm)	2	3	4	5	5.5
Height (cm)	11	6.5	4	2.3	1.6
Volume (cm³)	138	184	201	181	152

RESEARCH PHASE 4: SQUARE BOXES

3. Box D

    The width and depth of the base is twice the height.

4. Make the width and depth of the base twice the height. Or, the height is half the length and width of the base.

    Optional: Here is a more in-depth explanation that uses the area formulas to calculate what the dimensions should be when maximizing volume given a certain surface area.

    If the surface area is fixed to some value x and the box has a square base,

    x = AREA OF BASE + 4 TIMES THE AREA OF ONE SIDE

    (All sides except the base have the same area because the square base makes the width and depth the same. Demonstrate this using one of the containers, if needed. The height is the same for all four sides.)

    Let y = length (same as width) of one side of base

    $x = (y \cdot y) + 4 (y \cdot h)$

    Let $y/2$ = height (if the width and length of the base is twice the height, then the height is ½ the width and length of the base)

    $x = y^2 + 4(y \cdot y/2)$
    $x = y^2 + 4y^2/2$
    $x = y^2 + 2y^2$
    $x = 3y^2$
    $\sqrt{(x / 3)} = y$

    So, to maximize volume, make each base length the square root of the surface area divided by 3, and make the height ½ that value.

5. Both containers hold the most when the length and width of the base is twice the height. In the cylinder, we find that the container holds the most for a given surface area when the radius equals height. Another way of putting this is that the diameter is twice the height, since diameter is 2 times the radius. (Use a drawing to illustrate this, if it's unclear.) The diameter of the base corresponds to the base length of the square box.

6. Students should come up with something like the width (perhaps the diagonal across opposite corners of the base) is twice as long as the height.

ANSWER KEY

3. BRAINSTORM POSSSIBLE SOLUTIONS

1. 1987 cm²

2. Sketches will vary.

DESIGN CHALLENGE 3: BALANCING ACT!
2. RESEARCH THE PROBLEM
RESEARCH PHASE 1: BALANCING TWO OBJECTS

1.

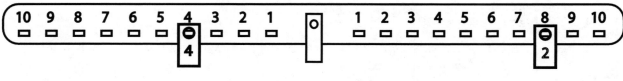

$$4 \cdot 4 = 16 = 2 \cdot 8$$

2.

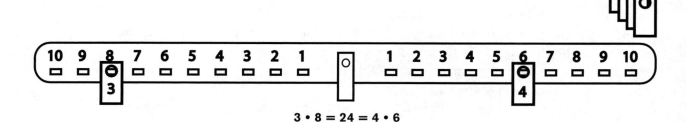

$$3 \cdot 8 = 24 = 4 \cdot 6$$

3.

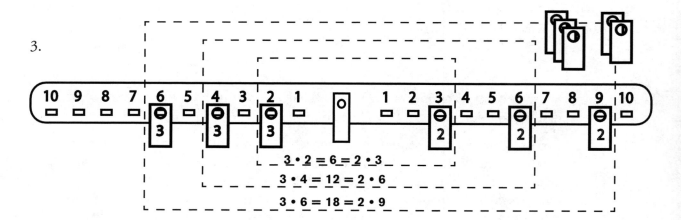

$$3 \cdot 2 = 6 = 2 \cdot 3$$
$$3 \cdot 4 = 12 = 2 \cdot 6$$
$$3 \cdot 6 = 18 = 2 \cdot 9$$

4. Answers will vary.

ANSWER KEY

RESEARCH PHASE 2: BALANCING MORE THAN TWO OBJECTS

1.

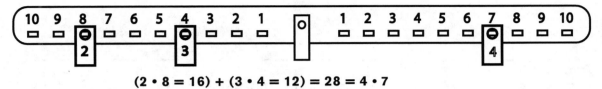

$(2 \cdot 8 = 16) + (3 \cdot 4 = 12) = 28 = 4 \cdot 7$

2.

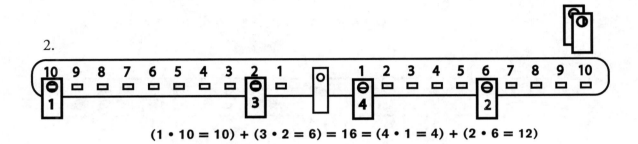

$(1 \cdot 10 = 10) + (3 \cdot 2 = 6) = 16 = (4 \cdot 1 = 4) + (2 \cdot 6 = 12)$

3. There are six possible solutions; one is shown below.

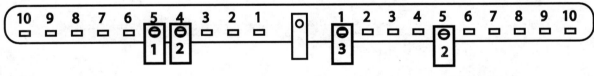

$(2 \cdot 4 = 8) + (1 \cdot 5 = 5) = 13 = (3 \cdot 1 = 3) + (2 \cdot 5 = 10)$

4. Answers will vary.

5. There are many answers; two possible solutions are shown below.

Solution 1

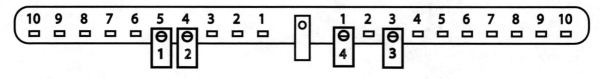

$$(1 \cdot 5 = 5) + (2 \cdot 4 = 8) = 13 = (4 \cdot 1 = 4) + (3 \cdot 3 = 9)$$

Solution 2

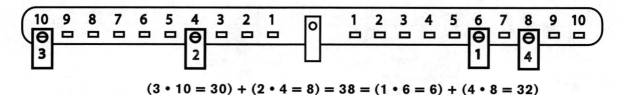

$$(3 \cdot 10 = 30) + (2 \cdot 4 = 8) = 38 = (1 \cdot 6 = 6) + (4 \cdot 8 = 32)$$

6. Answers will vary.

7. The sum of the products of weight • distance on the left side of the fulcrum is equal to the sum of the products of weight • distance on the right side of the fulcrum.

8. Let W = weight; let D = distance.

$$W_{L1} \cdot D_{L1} + W_{L2} \cdot D_{L2} + W_{L3} \cdot D_{L3} + \ldots = W_{R1} \cdot D_{R1} + W_{R2} \cdot D_{R2} + W_{R3} \cdot D_{R3} + \ldots$$

$$\text{or, } \Sigma W_L \cdot D_L = \Sigma W_R \cdot D_R$$

ANSWER KEY

3. BRAINSTORM POSSIBLE SOLUTIONS

1. 3
2. 4
3. 1
4. 1
5. 2
6. 2
7. 1
8. Answers will vary.

9.

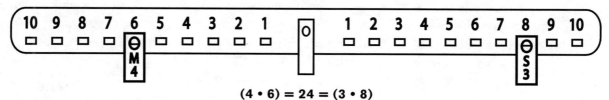

$(4 \cdot 6) = 24 = (3 \cdot 8)$

10.

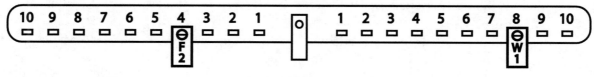

$(2 \cdot 4) = 8 = (1 \cdot 8)$

11. Answers will vary.

4. CHOOSE THE BEST SOLUTION

There are many possible solutions to this problem.

One possible solution:

When Σ(weight • distance) is calculated for this solution, right = left = 125.

Step 1: $M4 \cdot 6 = \mathbf{24} = S3 \cdot 8$

Step 2: $F2 \cdot 5 = \mathbf{10} = W1 \cdot 10$

Step 3: $S3 \cdot 4 + P1 \cdot 4 = \mathbf{16} = M4 \cdot 4$

Step 4: $R2 \cdot 10 = \mathbf{20} = S3 \cdot 5 + P1 \cdot 5$

Step 5: $S3 \cdot 7 + P1 \cdot 7 = \mathbf{28} = M4 \cdot 7$

Step 6: $W1 \cdot 9 + L1 \cdot 9 = \mathbf{18} = S3 \cdot 6$

Step 7: $S3 \cdot 2 = \mathbf{6} = S3 \cdot 2$

Step 8: $S3 \cdot 1 = \mathbf{3} = S3 \cdot 1$

ANSWER KEY

The numbers below the tiles indicate the order in which pairs of items are loaded onto the canoe.

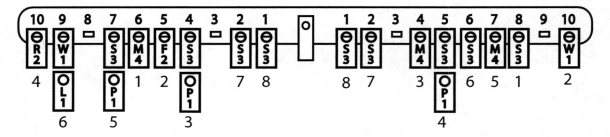

STUDENT WORK SAMPLE 1

1. 2
2. Some materials are labeled and purposes are indicated; also, although not all logs are drawn, the diagram does show a label of "17 logs"—presumably referring to the craft sticks.
3. Include more labels of materials and their purpose (e.g., clay, string). Include the scale chosen; include dimensions and units.

STUDENT WORK SAMPLE 2

1. 2
2. The drawing shows some dimensions that appear to be to scale, but it doesn't show scale. Also, it is unclear whether the student meant to use "ft" or "in," as both are included. Some materials are labeled, but it is unclear how materials are joined or for what purpose.
3. Label more materials (popsicle sticks or "wood"). Explain or show more clearly how rope is used. Include a scale. Show another view of the shelter.

STUDENT WORK SAMPLE 3

1. 2
2. The multiple views of the structure are helpful to show the shape of the shelter. However, the bottom drawing shows a pointy top while the top drawing shows a truncated triangle, so the design is somewhat ambiguous. Dimensions are shown, but no scale is indicated.
3. Label the materials used.

STUDENT WORK SAMPLE 4

1. 1
2. The drawing is missing dimensions, scale, and labels of materials used. But it does show some feasible ideas for a shelter (it has a roof, and lines could indicate the use of craft sticks).
3. Label the materials used, include dimensions and a scale, and express some rationale for the use of materials.

ANSWER KEY

STUDENT WORK SAMPLE 5

1. 2
2. The drawing includes labels for all items and people that need to be loaded onto the canoe, and clearly shows their locations on the canoe. However, it doesn't show mathematical justification or order of loading.
3. Number the items/people in order of loading and show how the canoe stays balanced mathematically.

STUDENT WORK SAMPLE 6

1. 2
2. The drawing shows all items and people that need to be loaded onto the canoe and their locations. It also shows the "weight" of each item and person on the canoe.
3. Number the order of the people and items loaded onto the canoe. Show a mathematical justification that the canoe will stay balanced.

STUDENT WORK SAMPLE 7

1. 1
2. The key is used to show some of the items and people loaded onto the canoe, but the list is incomplete. Also, it's clear that the canoe will not stay balanced.
3. Make sure that all people and items are loaded onto the canoe. Number the order of each pair of people and items loaded onto the canoe. Show mathematical justification that the canoe will stay balanced.

STUDENT WORK SAMPLE 8

1. 3
2. Drawing includes labels of all items and people, and shows the order that they're loaded onto the canoe. It also shows the "weight" of each item and person, and each item or person's location on the canoe.
3. Show a mathematical justification that the canoe will stay balanced.